PENGUIN BOOKS

Diddly Squat

'Til the Cows Come Home

Jeremy Clarkson began his writing career on the *Rotherham Advertiser*. Since then he has written for the *Sun*, the *Sunday Times*, the *Rochdale Observer*, the *Wolverhampton Express & Star*, all of the Associated Kent Newspapers and *Lincolnshire Life*. He was, for many years, the tallest person on television. He now lives on Diddly Squat Farm in Oxfordshire, where he is learning to become a farmer.

By the same author

Motorworld

Jeremy Clarkson's Hot 100

Jeremy Clarkson's Planet Dagenham

Born to Be Riled

Clarkson on Cars

The World According to Clarkson

I Know You Got Soul

And Another Thing

Don't Stop Me Now

For Crying Out Loud!

Driven to Distraction

How Hard Can It Be?

Round the Bend

Is It Really Too Much To Ask?

What Could Possibly Go Wrong?

As I Was Saying . . .

If You'd Just Let Me Finish!

Really?

Can You Make This Thing Go Faster?

Diddly Squat

Diddly Squat

'Til the Cows Come Home

JEREMY CLARKSON

PENGUIN BOOKS

PENGUIN BOOKS

UK | USA | Canada | Ireland | Australia
India | New Zealand | South Africa

Penguin Books is part of the Penguin Random House group of companies
whose addresses can be found at global.penguinrandomhouse.com

Penguin
Random House
UK

First published by Penguin Michael Joseph 2022
Published in Penguin Books 2023
001

Copyright © Jeremy Clarkson, 2022
Illustrations © copyright Garry Walton at Meiklejohn, 2022

The moral right of the author has been asserted

Typeset by Jouve (UK), Milton Keynes
Printed and bound in Great Britain by Clays Ltd, Elcograf S.p.A.

The authorized representative in the EEA is Penguin Random House Ireland,
Morrison Chambers, 32 Nassau Street, Dublin D02 YH68

A CIP catalogue record for this book is available from the British Library

ISBN: 978-1-405-95463-1

www.greenpenguin.co.uk

This book is dedicated to all of the world's farmers.

Contents

CONTENTS

The contents of this book first appeared in Jeremy Clarkson's *Sunday Times* column. Read more about the world according to Clarkson every week in the *Sunday Times*.

Introduction

Hello . . . in my first book about life at Diddly Squat Farm, I explained how I'd survived a year of biblical weather, a forest of red tape, a lockdown, a herd of escapey sheep, ten million vindictive bees, chickens, trout and supply chain issues. So it felt a little unsatisfactory to finish with a story about how I stopped a school trip from urinating in my garden.

. . . So, picking up where we left off . . . Diddly Squat's still going strong(ish). I continue to love what I'm doing even though I'm not very good at it. And to make things worse, I now have that most dangerous of things: a little knowledge. For instance, I know the difference between wheat and barley, and I know what oilseed rape is for. But what I know most of all is that farming is unbelievably hard work, and not very well paid.

There are times though, when a spare moment presents itself and the sun is shining and I can lean on a gate, chewing on a piece of grass. And then . . . it's the best job in the world.

SUMMER

Cows on the rampage

Last month I watched some alarming video footage of a policeman near Reading deliberately ramming a cow with his pick-up truck. The poor thing had escaped from its field and was wandering around on a road, in some distress. Apparently, it had knocked a shopper over and there were fears it could eat someone's lawn. But even so there was no need to kill it. It's not like it was a brontosaurus or a saltwater crocodile. It was just a cow.

So why didn't plod entice it with a bit of hay, or call a vet, who could have pulled something humane and scientific from his bag of Herriot trickery. Ramming it with a two-ton police truck seems so brutal.

We are told that the cow represented a significant danger to motorists but as the road had been closed this seems unlikely. What seems more likely is that whoever decided to implement a new policy of ramming cows is 'a bit thick'.

Still, the British plod did better than their counterparts in America because back in March, when officers in Virginia were asked to deal with an escaped cow, they reacted by arriving on the scene in a blizzard of noise and flashing lights and tried to shoot it – which resulted in them missing ten square feet of slow-moving cow and hitting one of their colleagues. Then, after he'd been taken to hospital and the cow had been wrangled to the ground by someone who knew what he was doing, they shot it anyway.

It's all too tragic for words but at least there is a silver lining. I've decided, against all my instincts, to install some new fencing.

When I was younger I used to sit watching the most interminable drivel on television because I couldn't be bothered to get out of the chair and change channels. Now, of course, thanks to the remote control I don't have to. As soon as James May says 'Hello and welcome to . . .' I just press the button and, kerpow, he's gone.

Nor when I want to speak with a friend do I need to go to the phone box and stand in a puddle of tramp urine, or wait for the neighbour to get off the 'party line' because today it's considered a 'human right' to have a smartphone of your own.

Life is easier and faster in every way. We don't have to crank a handle to start the car. We don't have to put up with feeling green around the gills for 90 minutes when we want to cross the Channel. We don't even have to go to the shops when we want something.

However, when we want to build a fence, we are still expected to use the exact same system as those apes we saw at the beginning of Kubrick's space odyssey. You deploy something called a 'fence knocker', which is a heavy bit of iron with handles on the side and is far and away the most tiring thing in the world.

Knocking in one post takes about 20 minutes and in a mile of fencing there are 1,320 posts. And on my farm there is maybe 20 miles of fencing. Get that job done and I could beat Tyson Fury at arm wrestling. That's why I use a different system, which is called 'paying someone else to do it'.

Except now, thanks to global warming, or transgender issues, or some other kind of woke plague, I can't even do that because there's a massive planetwide shortage of timber.

A small beetle, no bigger than a grain of rice, decided recently to eat Canada. It started in British Columbia, stripping the bark off every tree it could find, and now it's halfway across Alberta. God knows how it's still hungry after all of that but it is and soon there will be no trees left.

And it's not just the six-legged Mr Creosote that's causing the problem, because there's also the Wuhan bat. When that caused the world to shut down, some people went home and spent their newly gifted free time making bread. Some learnt French and some broke out the watercolours. But the vast majority reckoned that because they were stuck at home they may as well spruce the place up a bit.

So off they went to the DIY store, where they bought timber for decking, timber for new fencing and timber for an extension. And this wasn't just happening in Britain. It was happening everywhere.

So now there's a huge surge in demand for what the beetle hasn't eaten and, to create the perfect storm, there's also a shortage of sawmills. After the last recession in 2008, a lot of lumber yards suffered a kneejerk reaction to the sub-prime problems and shut up shop. So now, when timber is cut, there are very few places that can turn it into lumber.

And Covid has also caused chaos with shipping. So even if you can find a tree that hasn't been eaten, and you can find a sawmill that's still open, you're going to struggle to find a company that can deliver it. Which is why you might want to look at getting your wood from Europe. Yeah, well, post Brexit, good luck with that.

The upshot is that fence posts now cost more than most space rockets. And to buy one you have to meet a shady chap called Derek round the back of the pub.

Sure, I have planted many trees over the past year or so, but it'll be 30 years before they can be used to separate farm animals from idiotic policemen, which is why I turned my attention to fencing made from iron.

And it's the same story. Because wood now costs more than heroin, everyone has thought the same thing, and as a result every company making metal fences has filled its web pages with words like 'Due to unprecedented demand'. Which is corporate speak for 'oops, we've sold out'.

Walls are not the answer either. Yes, there's no shortage of stone. We are standing on it. But there is a shortage of people who can arrange it so that it doesn't fall over. I have a great guy on the farm but he's 72 years old and can barely keep up with the badgers that knock over what he repairs every night. Asking him to build 20 miles of wall would finish him off. And anyway a drystone wall is no real barrier to livestock. Sheep just clamber over it and, given the right motivation, a cow could win the Grand National. Pigs, meanwhile, are so good at escaping, they make Steve McQueen look like a rank amateur.

I'm fearful then that in the coming months we will see a lot more livestock on our roads and that maybe some new legislation may be necessary. Legislation that reminds officers they may not deliberately run down cows, or dogs, or any animal for that matter. Just burglars.

Growing chillies has cost me a fortune

Farmers have been told by the eight-year-olds at the Department of Agriculture that they must learn to stand on their own two feet. That they can no longer rely on the government for financial assistance. And that if they wish to survive they must do diversity.

That's why, over the past 18 months, I have peppered the farm with many new and exciting money-spinning projects, and then watched in horror as almost all of them ended in expensive failure.

My first idea was potatoes and they grew well. Sadly, however, I ended up with 16 tons of the damn things, which was not a large enough amount to interest the supermarkets, but it was too much to sell at the side of the road. And by the time I'd built a farm shop in which they could be sold, they had all rotted.

Next up there was my plan to rear trouts. I made a pond, provided shade so they didn't get sunburnt, and imported from America an automatic fish feeder. I sold a few to my local pub and ate a couple one night with some of the potatoes that hadn't rotted. And the rest were stolen by otters and herons.

Wasabi? This was by far my biggest disaster. I noted the fact that only one chap in all of Britain was growing it and saw this as an opportunity rather than a warning that it might be impossible. It isn't, if you really pay attention, round the clock, but I couldn't do that because I had potatoes to sell and fish to save and wheat and barley and rape to grow, and *Who Wants to Be a Millionaire?* to host and *The Grand Tour* to write, and when I did get five minutes to go and check on the 'green gold', I remembered that I had to write a newspaper column.

As a result most of the wasabi was eaten by my pheasants, which then ran around squawking a lot. Have you ever seen a pheasant sweat? I have. A few plants survived and I tried to sell them to restaurants in London, but this plan failed, and when I put them in my shop most of the customers took one look and bought a sausage instead.

In fact it turns out that no one wants to buy vegetables at all, which is a good thing because the five-acre plot I set aside to grow broad beans and spinach and so on produced six weedy sticks of something called chard. And that's about it.

It wasn't all bad, though. The bees worked hard and produced a lot of honey and the shop is proving to be something of a success – even though I'll have to close it down this winter to give it the fake slate roof that the local council likes, instead of the tasteful green roof it has now, which everyone else likes.

Anyway, because I have tasted the sweet smell of success in two of the 400 projects I started, I've decided to keep going and grow chillies.

This meant I needed a couple of polytunnels. We see a lot of these in the countryside at the moment and certainly I couldn't see a problem putting them up as I once erected a tent, in a gale, at the North Pole, with nothing but James May on hand to help. I'm a seasoned tent builder, me.

I suspected I had bitten off more than I could chew, however, when they arrived in a large lorry and had to be manhandled to the ground by my telehandler. And then my fears were confirmed when I found the step-by-step instruction book, which is 60 pages long and starts with a warning about not attempting to do the job on your own.

This meant I had to consult my new gardener, who began by saying that first of all I needed to build some French drains for irrigation. And that the site I'd chosen was too exposed so I'd need to build a protective bund.

There is some debate about whether you need planning permission for such a thing, which meant I didn't bother asking. Instead I took the spoil from the building site that is my new house, smothered it in top soil and sprinkled that with wild flower seed. It's already growing well and is home to many insects that will be put on to the street when the council comes round with one of its endless enforcement notices.

The drainage system is in too. It only took three men four days to complete, but now the wind and the rain can do their worst and my chilli-filled polytunnels will be unaffected.

First, though, I had to build them, so I sat down with the 60-page booklet and realised by the time I was half-way down page 1 that I didn't have the patience to go any further. All the diagrams looked exactly the same to me and there were words like 'swaged' that I didn't understand. I realise now that this was explained at the bottom of page 2, but I didn't get that far before dragging Kaleb from his tractor and asking him to help out (do it).

Pretty soon Kaleb was cross with me because in my experience tape measures don't give accurate readings. I figure, because it's what I'd do if I worked in the factory where they were made, all the markings are slightly different.

Kaleb said this was rubbish and stomped off to measure out where the foundation poles should be located.

Then we had an argument about whether foundation plates would be needed as well, which resulted in a call to the manufacturers, which resulted in us discovering that we would need to mount the foundations in concrete.

So out came the digger again and out came the Postcrete and after a week we had the framework of one tent up. It was horrific work, filled with manual labour, sunburn, argument and hay fever. Work began this morning on the second tent but, sadly, I'm inside writing this so I can't help. Tomorrow, though, after I've launched the latest *Grand Tour* in London and written another column, I'll probably definitely be back out there, holding spirit levels and operating spades.

With or without me, the polythene fabric will soon be stretched over the framework, the sprinkler system will be initiated and my chillies will be in there, turning sunlight, manure and warmth into little pods of fiery goodness.

I was tempted to go for the hottest variety possible but was steered from this course when I learnt that they are extremely difficult to grow. And there's always the worry that someone will be disinclined to buy a second jar of chilli chutney if the first one turned the bottom half of their face into a ruined bloody goo.

So I'm going for medium-hot jalapeños, which grow

well and taste good. The only problem is that by the time the polytunnels are up, I'll be about £20,000 down. Which means that to turn a profit my chutney will have to be about £500 a jar. The jam could be a bit more.

So. Diversity. Sounds great in principle of course. But it rarely works in practice.

No one can destroy a plant like I can

For the past 18 months my new house has been emerging from the ground in buildery slow motion, but it now has a roof and windows so it's time to start thinking about what sort of garden I'd like. Which is why I was delighted to receive an invitation to last week's preview day at the Royal Horticultural Show at Hampton Court.

I'm not much good at lots of things, but I'm especially not much good with plants. Everything I'm ever given dies because I water it too much or not enough or I put it in the wrong room and either too near a radiator or not near enough. I think I'm the only person in all of recorded history to kill a spider plant.

Outside, things aren't much better. I planted a wisteria last year in a stone trough and then found out, after it died, that you can't plant a wisteria in a stone trough. My clematis melted. My new trees gave themselves a Brazilian. And after I sprinkled some seeds in the garden of the cottage where I've been living, I ended up with what looks like a municipal roundabout.

I can't even operate gardening tools properly. It doesn't

matter how German or robust they may be: I bring them home from the shop, they snap immediately and then I have to go back to the shop to buy replacements. Last year I spent nearly £4 million on trowels.

But I was looking forward to what I figured would be a genteel day at Hampton Court, mingling with agreeable Volvo people in elasticated Monty Don action trousers. And hopeful that I'd come away with an understanding of what to buy and where it should go.

The centrepiece of this year's event is a crashed airliner with the words 'homo sapiens' emblazoned down the ruined fuselage. It sits on a bed of charcoal, gently smouldering, amid a pile of suitcase and seat-cushion flotsam in a field of wheat.

As my girlfriend was once in a Learjet that skidded off the end of the runway at Northolt airport and onto the A40, where it was cut in half by a van, she was a bit disturbed by this. Me? I was bemused. Because why would anyone want a crashed airliner in their garden? It seems a silly idea.

It then transpired that this is supposed to remind visitors about the planetary consequences of air travel, but again I was bemused, because almost no one has been on a plane for 18 months. And most won't be going on one this summer either.

Having decided that the organisers had obviously

taken leave of their senses, we moved on to the next exhibit, which was a Vietnamese street food stall selling pho, where I was accosted by a woman with unusual hair and a dirty dress who said she was a journalist from the *Telegraph* or the *Mail*. She wasn't sure which one.

Anyway, while I was trying to order my pho ga, she asked what I thought about the ecological theme at the show this year. I pointed out that it's hard to have a garden show that doesn't have an ecological theme unless you put a crashed airliner in it. This seemed to make her a bit exasperated. So she said that most gardens involved 'killing things', and then off she went to write an idiotic piece saying I was once sacked by the BBC and that I spray everything in my garden with glyphosate. Not sure she's fully up to speed with recent events.

After this encounter and my rather good pho, I went to look at stalls selling cufflinks and shoes and fountains until, eventually, having found no ecological theme at all, and no plants, I found a patch of lawn full of Anthea Turner and Tony Robinson.

In a nearby marquee, however, there were some bonsai trees that I liked, and some flowers, all of which came with labelling in Latin. This is not a language in which I'm proficient. Like all good public school boys, I can conjugate a table, but I certainly wouldn't feel comfortable using it to create a garden. That's why, when a Monty

Don trug person came over to ask whether the new Land Rover Defender was any good, I replied in German. So he'd know how I felt.

Many of the plants I liked were Japanese, so I felt unconfident they would grow well on the top of a hill in the Cotswolds. I had similar doubts about some squiggly ones from South America, dramatic though they were. I then asked a man about some moss that looked interesting. '*Ego bet te qui oderunt vivamus ingruat*,' he said, with a smile and a wink.

I'd gone to Hampton Court because I'm trying to create a sunken garden, an orchard and some formal flowerbeds at the side of the house, but after two hours I'd not found anything that would look remotely at home. I was so fed up, I was even entertaining the idea of deliberately annoying the weird woman from the *Mail* or the *Telegraph* by coming home and concreting over the whole thing. And then decorating it with some dead badgers.

But I persevered until, eventually, I found a nice man who was selling climbing bluebells, or *Sollya heterophylla* as he called them. He assured me that the plant would grow in the shade or the sunshine and in any kind of soil *luto absque*, and so I bought one for 25 silver denarii and brought it home. Then, this morning, I found it in the boot of the car, on its side, snapped.

Another thing I came home with, apart from a dead plant, was a high temperature and some breathlessness. Which is annoying, because I have tickets for both Wimbledon and Wembley today.

Hopefully I will be Covid-negative, as I'm looking forward to watching the players trying to steer their balls round the crashed airliner some damn fool has put in the middle of the playing area to make some kind of point about something or other.

The animal welfare state

You never hear anyone say, 'I hate elephants and lions and cows and I go to sleep every night dreaming of the day when I can pull the legs off an otter.' This is because everyone loves animals. We like looking at them. We like petting them. And we like eating them.

Right. So one day you're in the supermarket and in front of you are two legs of lamb. One is from the UK and costs £20 and one is from New Zealand and costs £15. So that's an easy choice. You buy the one from down under. Lovely.

But it isn't lovely, because animals farmed in New Zealand and America and China and Brazil and Canada and Australia – with which Boris has just done a much-trumpeted trade deal – do not have anything like the happy lives enjoyed by the animals farmed here.

Yes, it's true, my boy sheeps had a rubber band placed around their scrotum so that after a couple of weeks their testicles would fall off. That sounds like a terrible thing to do, but it stops them from getting jiggy with their sisters. Or indeed their brothers.

Their tails are also removed because leaving them in place might cause blowfly strike. As it's a Sunday morning and you're probably having breakfast, I won't go into the details, save to say that being eaten alive by maggots is one of the more revolting ways for an animal to die.

In short, sheep reared here, and cows and pigs and all the delicious things we eat, are treated with one eye very firmly wedged on their welfare and to hell with the costs involved. That is emphatically not the case with many of the animals reared abroad.

In various US states, for instance, female pigs are placed in 2ft-wide cages for 16 weeks when they're pregnant. This stops them from fighting and attacking or even eating one another and therefore increases profitability. Excellent news for money-minded shoppers. Not so good for the pigs.

Still, could be worse. In the Philippines pigs have been doped on something called ractopamine, which increases the amount of meat on the animal significantly but means that many could win the Tour de France. On the downside it causes some of them to tremble so violently that they can break their own legs.

I'm thinking of getting some pigs on my farm. But instead of cages and Lance Armstrong medication, I was thinking of putting them in my woods, where they can snuffle around and live on a diet of wild garlic and watercress. I don't doubt the meat they produce will be more expensive than the alternative from Russia or Canada, but if shoppers could be educated . . . ?

I get why Boris did that trade deal with Australia. He'd led us out of the EU and needed to demonstrate that it was entirely possible to do international business on our own. That's why he stood there, on his PR war footing, saying that people in Sydney could now enjoy some good old British Marmite. I seriously doubt that but I can see where he's coming from.

And now it's time for me to retort. Because the deal he did is a two-way street. They get our Marmite and we get several million tonnes of their beef, which comes from cows that were hot branded, filled with hormones and then slaughtered in unmonitored abattoirs having been transported for hours in temperatures that could melt rock.

I've been to an Australian cattle farm and it was truly a sight to boggle the mind. Covering an area bigger than Kent, Sussex, Hampshire, London, Essex and Hertfordshire – that's one farm – 40,000 cows were to be found mooching around in what appeared to be a big red desert. God knows what they were eating. Or doing to amuse themselves. It was mass production, a gigantic outdoor factory, and the meat it produces will soon be coming to rival the cottage industry that is beef farming in the UK.

Naturally farmers here feel helpless. They are forced by the law to abide by UK animal welfare rules, and then they are told they must compete on price with farmers in other countries who are not. It's impossible.

Solutions? Well, they could take the government shilling and, instead of farming, move to a bungalow in Worthing so that their land could be turned into a Lycra-based theme park for extreme cyclists and disciples of Street-Porter and Packham. Or they could fight back and try pointing out to British consumers that food from abroad is cheaper because it's shit.

This is a tricky message to get across, because after 20 years of nuking our taste buds with bread that's mostly sugar, Ronald McDonald's special sauce, chicken vindaloo, deep-fried chicken and crisps made from artificially flavoured carpet underlay, most of us could not tell a beautiful piece of prime beef from a Walnut Whip.

Like most middle-class people, I sit in restaurants these days listening to the preposterous waiter explaining how the wagyu beef was raised and how long it was hung for and as a result I think, as I eat it, that it was obviously worth £450,000. But the truth is that in a blindfold test I couldn't tell it apart from the slab of fly-bitten gristle I once had in Chad.

So there's no point trying to explain that British meat tastes better than Australian meat or New Zealand meat or American meat because, even if it did, you'd need to be a beef sommelier to notice.

What farmers here can say, though, is that the animal they raised will have been treated more kindly before it ended up on your plate. I believe that will strike a chord.

Let's show consumers pictures of chained veals in America, lapping away at what really doesn't pass muster as milk. Let's show Brazilian cattle being given drugs until they look like Arnold Schwarzenegger, and tell them that the antibiotic levels found in the US dairy industry are eight times higher than the permitted levels in the UK.

I haven't finished. Let them see pictures of adult sheep in Australia being castrated without anaesthetic and live cows being transported on ships in tropical storms. You've been seasick, I'm sure; it's awful. Well, now imagine that sense of desperation in an environment where you can't lie down, you have open sores on your face, the water you've been given to drink is contaminated and you are coated in a thick layer of your mate's faeces.

Of course, there's always the worry that if we do this people will decide immediately to become vegetarians. So to counter that we need also to show them pictures of life for farm animals in the UK. Lush grass. Babbling brooks. Refreshing showers, interspersed with longer periods of rain. Kindly farmers. Well-run slaughterhouses. Cows fertilising fields with their dung. And we need to show them that the researchers who told *The Sunday Times* recently that potatoes are better for the environment than cows is, to quote Cheerful Charlie from my farm programme, 'almost certainly just plain wrong'.

At the moment Carrie Johnson and her lapdog in No. 10, backed by Chris Packham, the lefties on social media, Greta Thunberg and the vegetarian movement are running a powerful anti-farming and anti-meat campaign. It feels, from where I'm sitting, like a tsunami of misinformation and stupidity.

So we need to fight back. We need to sit outside supermarkets with the food we grow, in our tractors, telling shoppers that if they really care about animals and they really care about nature and they really want to see red admiral butterflies in their garden, they'd better wise up and start eating British meat. Even if it does cost an extra five pounds.

My battle with a troublesome weed

In most jobs you tend to encounter the same sort of problems every day: the Wi-Fi router breaks, someone phones in sick, then the photocopier runs out of toner, again. Farming, however, is different, because every day you are faced with a problem you couldn't possibly have seen coming.

I listen to normal people moaning about how someone nicked their parking space in the office car park that morning and I think yes, annoying, but you haven't had to deal with a sheep that's climbed into a barrel of oil and drowned, or a sudden downpour that's totally ruined your entire crop of wheat, or a gremlin in the unfathomable PTO unit of your tractor.

For the past couple of weeks I have been watching, with some concern, a plague of greenness emerging in the barley field at the back of my cottage. So I called Cheerful Charlie, my land agent, who, naturally, said it was a 'big problem'.

It's called brome, apparently, and it's a weed that competes with the barley for nutrients and sunlight – and it usually wins. Which means you end up harvesting the square root of bugger all.

To begin with we had to work out where it had come from. Charlie reckoned the seeds that spawned it had been lodged in a drill we'd borrowed, and I was all set to insert something agricultural in Kaleb, my tractor driver, for not washing it out.

But just before the squeal-like-a-piggy scene began, Charlie had another thought. A year earlier I'd used the field to graze sheep and, he reckoned, it's possible they could have introduced the brome in their excrement. 'Bastard sheep,' I muttered. Even from beyond the grave they are ruining my life. But it wasn't the sheep either because on my afternoon walk I noticed that it wasn't just one field that had been infested. It was all of them.

Instead of sinking to my knees and weeping, which is what I was minded to do, I had to rush off to deal with other problems, like the Brummies who'd seen my farming show and were out for a walk in the middle of a wheat field. And the local who had written to the planning authorities saying I had never used my lambing barn for lambing. Plainly this is a man who hasn't seen my show.

When I got home for a quick cup of tea, a neighbouring farmer came round to say his lake was jammed full of annoying and invasive, but delicious, American crayfish, which I could perhaps sell in the shop. That all sounded very simple – eating what is effectively an aquatic grey squirrel – but, of course, the government won't let me catch them without a licence.

And then Kaleb dropped by to say that after they'd been caught, they'd need keeping in a tank for three weeks and fed on a diet of potatoes before I could sell them as a locally produced snack to people who were tired and hungry after a long walk through my wheat. Like I say, the problems farmers face are never simple and you never see them coming.

Anyway, I got talking to the man with a lake full of American crayfish and he said that his barley had also been affected by brome, which meant we stopped trying to work out how it had got there and moved on to the issue of how we'd make it go away.

My neighbour had nuked virtually his entire field. 'You have to,' he said, 'or you'll be plagued by it for ever.'

Kaleb reckoned this wouldn't be necessary as we will be growing oil-seed rape in the barley fields next year, and that, for reasons I couldn't understand, will somehow deal with it.

And that's the other thing I've learnt about farming. There's never a right or a wrong way of sorting something out. Even after 12,000 years of agriculture we are still feeling our way around, sucking things and seeing.

According to Charlie, the easiest way of dealing with brome is to burn off the stubble, like we used to in the olden days. But we can't do that any more because of the smoke it creates. And his answer to my next question was: 'No, Jeremy. You cannot have an accidental fire either.'

Later that night I went on the internet, where I discovered that ploughing a field will bury the brome seed, causing it to suffocate. But we can't do that either because dragging a plough through the soil means chewing through millions of gallons of diesel, and it releases carbon from the soil. So it's a climate change double whammy.

I could cultivate the fields after harvest and then use glyphosate, but this is unlikely to be a long-term solution as glyphosate is becoming a dirty word. There's even talk of it being banned and that's keeping me awake at night.

I can't burn the brome, I can't plough it into the ground, and soon I won't be able to kill it using chemicals. From an armchair in Whitehall all of this makes sense. We need to address climate change and we need to protect insects. But where does that leave me? Sitting here, looking at a field of weeds, that's where. But with the germ of an idea.

As we know, proper shops in town centres are being replaced with idiotic juice bars, where socialists compete with one another to use the daftest, laburnum and nettle 'I saw you coming' ingredients to win the hearts and minds of thin urban women who like to start their day with a glass of green slime.

Well, what if one of those ingredients was brome grass? Of course, we know that the human digestive system is not designed to deal with ordinary grass – that's why cows have four stomachs – but it turns out that brome is technically edible. Not the fibre, but the rest of it.

I don't doubt that it will have absolutely no nutritional value and will taste worse than a slab of marzipan-flavoured Marmite, but this is of no consequence to the young and the thin and the foolish. Get a couple of Instagram influencers on board with the idea that brome is the new eco-friendly alternative to avocado and we are away. I may even throw in some black grass as well to give it a bit of, oh, I don't know, 'sustainability'.

I guess this is how farmers must think from now on. To deal with the hot dry summers that we are told will be the new normal, we stop growing milling wheat and grow durum wheat instead, which is used to make pasta. I've tried it this year and it's doing quite well, by which I mean 'not disastrously'. And instead of worrying about brome and black grass, we monetise it. I already have a name for mine: Jeremy and the Juice.

When it becomes successful I shall need a head office and a car park, and every day people will phone in sick and the photocopier will run out of toner. And I shall enjoy this. Dealing with problems that you can see coming and which, as a result, aren't really problems at all.

Learning to moan like a farmer

When I began this farming malarkey back in 2019, we had one of the wettest autumns on record and I struggled to get my seeds into the ground. This year I'm struggling to get them out.

Two weeks ago I started the harvest. And it still isn't finished. And if the weather forecast is to be believed, it won't be finished for another ten days, by which time the rape seeds will have fallen out of their pods and the barley will have drooped to the point where it's unharvestable.

The figures suggest that this summer has not been especially wet, but I can assure you it has been wet at all the wrong times. We have a shower and we have to wait until the crop has dried out before we can fire up the combine.

So every two hours we do a moisture test and then, when the readings get into the acceptable range, we have another shower, and the cycle starts all over again. It's fantastically frustrating because there is nothing we can do about it. Apart from buy a barn with drying fans in the floor. But they're half a million quid, so I won't be doing that.

On one evening the barley was dry enough (just) and we got 50 acres done before the moisture meter started to suggest the dew was coming and it was time to stop for the night. Off went a full lorry load to the grain merchant, who called the next day to say that we hadn't stopped in time, and that as a result 1.2 tonnes of the 30 he'd picked up was basically water. Understandably he said he wouldn't be paying for that, so £170 was knocked off the price. And let's not forget, £170 is £26 more than the farm earned in total last year.

There was more too. Because the 29 tonnes of actual barley we'd sent him had to be dried before it could be turned into hen food, we'd be billed £256.

Still, at least things then got worse, because five of the tonnes we'd harvested wouldn't fit in the lorry and as it wasn't worth getting another truck for such a small amount, it's been sitting in the open for a week, becoming crusty and damp and useless. So that's another £700 down the drain.

Oh and because we still have half a big field of barley left to harvest, I can't get next year's oil-seed rape in the ground. Which means that when we do the dreaded flea beetle will be out there, knife and fork at the ready and its napkin tucked in. Still, I guess that if the rape is all devoured by beetles, at least I won't have to worry about the pigeons eating it.

What all of this moaning proves is that I'm starting to become a farmer. I can now whinge for hours, without repetition or hesitation. The weather. Defra. Carrie Johnson. That bloody alpaca. Chris Packham. Brexit. Badgers. Ramblers. The timber shortage. Flea beetles. Black-grass. Sheep. There's a bottomless pit of misery and despair in farming and I'm wallowing in it.

However, there's also some intense joy, which runs round my arterial route map like warm honey. And it's never warmer or more honeyish than at this time of year when I'm on the tractor and it's time to pull alongside the combine so that six or so tonnes of grain can be transferred from its innards into the trailer I'm towing.

Of course this could be done when both vehicles are stationary but that would be too easy, so it's done on the move. This is something that requires concentration. And I don't mean the sort of concentration the pilot of an Apache gunship must deploy when he's doing fighting. Because all he has to do is fly the helicopter, operate the chain gun and keep one eye on the target acquisition panel while trying not to be shot down. And that's nothing compared to the trickiness of loading a grain trailer.

First of all you must make absolutely sure that you

don't drive over the crops that haven't been harvested, or the line of straw that has. This means it's critical that you look forwards as you drive along. However, it is equally important that you are correctly positioned so the grain goes into the trailer and not back into the field from which it grew. This means it's critical that you look backwards as you drive along.

Ideally a farmer needs to have one eye on either side of his head, like a cod. And a neck made from horse because after five minutes of frantically looking forwards and backwards in a tractor that's bouncing over rough ground, mine felt like I'd just spent a day in an out-of-control g-force machine.

Then there's the business of loading the rear of the trailer first, because if you get a big mound at the front you won't be able to see over it when you have to load up the back. So you must drive at precisely the same speed as the combine and then slow down when you want to move the load forwards. It's counterintuitive and I'm very bad at it. Kaleb, my tractor driver, says he can't watch when I'm doing it and that I'm a . . . w . . . ell, let's say a lot of my seed falls on fallow ground and you'll get the gist of his observation.

Occasionally, though, it'll all go where it's supposed to go and then I must hurry back to the barn to tip it into a neat pile. This is hard for two reasons. First, I'm never quite sure which button to push to make the trailer go up. And second, I'm very bad at reversing. Six goes. That's my best effort yet to get it into the barn. Seventeen is my worst. And on that occasion I was still trying to get my load backed up when Kaleb turned up with his, and I had to pretend I'd stopped off for supper on the way. That seemed to make him even crosser. Apparently you don't stop for meals when you're harvesting. And nor – as he pointed out to me later – do you stop off at the farm shop for a nice cold pint.

Despite the difficulties and the concentration and the endless tellings-off, though, I love whizzing back and forth. I love the wildlife you see. I clocked some English partridges in one field and in another an albino fallow stag. And then after dark I could see the lights from all my neighbours' machinery as they rushed to get their harvest in before the rain as well. It all feels very important, somehow, to be making food.

All of us, though, were forced to stop by rising moisture at two in the morning and when I climbed between the sheets, all dusty and manly, I got a murmured 'How did it go?' from the other side of the bed. 'Very well,' I said before entering the land of nod.

The next morning, however, it seemed that it hadn't gone so well because somehow I'd knocked all three of the bins over, spilling rubbish everywhere and breaking them. And I'd completely destroyed a five-bar gate. I don't remember doing any of it. Literally I thought I'd got home without a scratch. Maybe that's why everyone is now calling me the Wolf of Chipping Norton.

But it's not the end of the world because it was a very happy day and now I have something new to moan about.

AUTUMN

The trouble with cows

As one of the world's leading and most influential eco-warriors, I've decided to try my hand at something called mob grazing. It's time-consuming and difficult, but as it's claimed to be the holy grail of environmentally friendly farming, I figured I'd channel my inner Attenborough and give it a bash.

Here's how it works. You fence off a small part of the field and put a flock of cows in there, so they eat every blade of grass and defecate on every square inch. Then you move them to the next patch and put hens in the bit of the field where the cows have just been. And you keep this going all summer. The hens eat the worms out of the cow shit, which gives them the protein they need, and as they cluck about they trample the cow faeces, and their own, into the soil, which revitalises it, meaning I don't have to use chemicals. It's all completely natural and to my mind brilliant.

The only tricky bit is that to make it all pay I have to turn the cows into meat one day and then sell it. Which is hard now because Boris – to try to prove that Brexit was a good idea – has done a deal with the Aussies that allows them to sell their beef in the UK. And somehow meat produced from grain-fed cows on the other side of the world will be cheaper in a Chipping Norton supermarket than the meat from cows that ate free grass here.

The only solution is to cut out the middleman and sell the meat myself, in my own restaurant. But plans to open one in my former lambing barn have gone down like a shower of sick with a few red-trouser people in my local village. Which is why I was to be found a couple of weeks ago hosting a cheese and wine town hall meeting with them. It was a (mostly) polite battle between the red-tractor movement and the red trousers and I think I did quite well. Certainly they didn't make a wicker man of me afterwards. Maybe that's why some people call me Boutros Boutros-Clarkson.

My next battle is with the planners and then I shall have to start work on the vegetablists, who have it in their heads that ruminants are somehow bad for the environment. Right. OK. So let's get rid of all those wildebeest on the Serengeti, shall we? And then we can start work on the giraffes. And Bambi. Or would it be better, I wonder, to shut up, buy British and enjoy a Sunday roast, knowing that you've made the soil on my farm fizz with life?

I'm getting ahead of myself here. Because first of all I needed some cows.

I was assured by various local chaps with checked shirts and faces like wizened walnuts that cows would be a thousand times easier to look after than sheep because they don't actively want to die. A sheep lives for the day when it can cut its own head off in a fence. Cows don't.

Mind you, there are some issues. I bought 20 but only 19 turned up because one, a bullock, had a testicle that hadn't dropped. This meant it couldn't be castrated and that meant it could very easily make its mother pregnant, or its sister.

It's exactly this kind of sexual liaison that produces government agricultural policy. I'm not kidding. Defra recently said that farmers must start using animal manure to breathe new life into the soil. But then you have the Environment Agency, which says farmers must not use animal manure to breathe new life into the soil in case it gets into the water supply.

They allow Thames Water to pump thousands of gallons of human sewage into the rivers of Oxfordshire, but I must not let my cows 'do their toilet' anywhere near any of the streams and springs on my farm. This meant I had to build two miles of fencing. And with timber now costing more than heroin, I ended the day with tears in my eyes, only some of which had been caused by the creosote.

I was also confused because in my mind there are cows, bulls and calves. However, it turns out this isn't so. Because there are also heifers and stores, and bullocks and steers, which for some reason are the same thing. All I know for sure is that my cows are shorthorns, which is odd because they have no horns. They are very pretty, though, and have hair like Kaleb's.

When they arrived they immediately set off to check the perimeter, much like we do when we arrive at a holiday hotel. Only, unlike us, they got to the bit where I'd used an electric fence instead of timber and the calves walked straight underneath it.

So we got them back in the field and off they went again until they reached an old bit of fencing, which they pushed over so they could inspect my game cover. And then, after we got them back to where they were supposed to be and we had mended the fence, they found their water trough. Which was broken. I'd bought a pump to get water from the stream to the trough, so that the Environment Agency wouldn't send me to prison, but somehow it wasn't working. So that was the rest of the day gone, trying to fix it. Cows, then, may be easier than sheep, but that doesn't make keeping them simple. It's not.

It's not even that safe. Shortly after evidence began to emerge that the AstraZeneca vaccine could cause blood clots, we were told that you'd stand more chance of being killed by a cow. This sounded very reassuring, but the truth is that five people a year are killed by cattle in the UK. This makes them more dangerous than motor racing. A lot more.

Mine have been here a few days now and so far I haven't been able to teach them to attack the ramblers. In fact I haven't been able to teach them anything at all. Because all they do is stare at me. They're like a pack of six-year-old kids when a stranger walks into their class-room. They're not inquisitive or interested or frightened. They just stare.

Soon, though, they will be able to stare at the hens as they eat the worms from their faeces and shortly after that they will stare at the workmen who are coming to build them a £100,000 barn to keep them warm in the winter.

If they make it that far. The problem is that badgers give cattle TB and there are hundreds on this farm. And I'm not allowed to shoot the damn things because of Brian May out of Queen.

That's where we are at with modern farming in Britain. I'm told by our esteemed leaders that to save the soil I must keep cows. But I'm then told that they must not urinate in the stream and that I must sit back and let the badgers give them a disease.

Well, if they do and the government sends round an execution squad, I shall simply tell them that while my cattle look like cows, they actually identify as alpacas. It'll be fun watching them work that one out.

In which I'm opening a restaurant . . .

Every morning, while my eggs are boiling, I turn to the obituaries section in *The Times* and console myself by noting that most of the people were 15 years older than me when they opened the door to find Mr Reaper standing there with a scythe and an apologetic demeanour.

This means I have 15 years left, and that's plenty. It took 15 years to get from my birth to my O-levels, and that felt like for ever.

So there's still time to cram in a lot more fun and games. Except, of course, there isn't, because the perception of time is not constant.

When you're 15 years old, 15 years is 100 per cent of your life, but when you're 60, it's 25 per cent. Which means that time, for me, now, is apparently moving four times as fast as it used to.

And it's speeding up even more with each passing day, so that soon it'll feel as though I'm on a full-bore warp drive to the edge of Klingon space.

There's another issue too, because what exactly am I going to do with the blink of time I have left? Become a kickboxing champion? Take up water-skiing? I don't think so. My knees give me no confidence when I'm coming down a flight of stairs. My back locks solid if I attempt to walk up a hill. My lungs feel as if they're on fire if I even look at a bicycle, and when I go for a swim it feels as though I have a small car on my back.

It saddens me to think that I have now dived off a boat for the very last time, and been down my last black run. I may never see the dawn again, either, unless I have to get up early for another annoying stagger to the loo.

And things aren't going to get any better, because soon there will be lumps and gristle and hip operations that will force me to spend what time I have left in a rocking chair, trying to finish an interesting story in the Reader's Digest about azaleas.

I've been lucky in life. I've seen the morning in the mountains of Alaska, I've seen the sunset in the east and in the west, I've sung the glory that was Rome and passed the hound-dog singer's home, and, unlike Noddy Holder, who's also done those things, I've looped the loop in an F-15 Strike Eagle, had tea with Nelson Mandela, roared through Phnom Penh in a jet-propelled speedboat and met Kristin Scott Thomas.

Despite the fact I've been busy, I don't want to admit I've finished. I still haven't written a work of fiction. Well, I have, but only by accident. I haven't done one on purpose, full of explosions and lantern-jawed people called Clint Thrust.

But what's the point of starting one now? I'll only get arthritis halfway through and be unable to type. And then, when I'm sent on the promotional tour, I'll stand there like Joe Biden running all my words into one and forgetting everyone's name.

It's the same story with my travel ambitions. I haven't been to Zimbabwe, and I'd love to go. But the flight is long, and I'll get heat rash, and I've seen a hyena already, so it's easy to think, 'Oh phooey. I'll just go to Cornwall instead.'

This, then, is the disaster that awaits us when we reach the final act. We don't have the time to do anything, but that's OK because we don't have the energy or the will either. That's why Billy Joel hasn't released a studio album since 1993. What's the point when the only people who'd benefit are his kids?

Right. So how come Donald Trump, at the age of 75, is making noises about going back to the White House? How come Patrick Stewart reprised his Captain Picard role at the age of 78? Why is Jilly Cooper still writing? Why is Mary Berry still cooking? Why is Harrison Ford still Indiana-ing? Why am I, at the age of 61, thinking of opening a restaurant? And why do I think everyone over the age of 60 should be thinking along the same lines?

Well, Britain is at present experiencing shortages of things because we don't have enough people to make stuff and pick it and move it about. So why not become a lorry driver? Or help us get round the mattress shortage by setting up a UK business collecting feathers, so we don't have to import them from China? Or make some shipping containers in your back yard, because how hard can it be?

Old people are very happy to sit about in their inconti-trousers, doing nothing all day, and then moaning about how young people are all too spoilt and entitled to get off their backsides.

Really? So what about Emma Raducanu or Kaleb, my young apprentice on the farm, who's basically an 18-hours-a-day Duracell bunny? Then there's the 14-year-old kid in my local village who makes eco dog biscuits on his mum's kitchen table, and my eldest daughter, who once took a day off. But no one can quite remember when it was.

I see lots of driven young people every day, but rarely do I see an older person charging round the place, all elbows and fire and steely determination. Which means there's a vast and experienced pool of talent going to waste.

And that brings me on to a plea. I have all the ingredients I need for my new restaurant but no clue how they might be turned into stuff that people might want to put in their mouths.

So is there anyone old out there who knows how to run such an establishment?

I hope so, because I want a kitchen full of pies and gravy and wipe-down chairs and Bad Company on the stereo and everyone exchanging bewildered looks when someone asks for the transgender lavatories.

It'd be fun, and that, as the clock ticks down to zero, is all we can hope for.

A new farmhouse by Christmas?

You would not build your own car or make your own trousers, so why on earth do people long for the day when they can build their own house? Oh, I'm sure you've watched *Grand Designs*, in which a couple take an hour, minus the ad breaks, to throw up a gleaming and sharp-edged masterpiece that cost them just £4.75 – but I now know from experience that the reality isn't like that at all.

I began work on my house nearly ten years ago. I contacted a respected architect, and we'd spend hours together toying with ideas and possibilities until, eventually, he produced an artist's impression of how it would look. And immediately a friend said: 'That's hideous.'

It's strange. I liked it. The architect liked it. And lots of other people who'd seen the design liked it.

But I couldn't go ahead and build something knowing that out there in the wide world there was one person who didn't like it. It would keep me awake at night, so I found another architect and we started again.

What I wanted was part Louisiana plantation house

and part Quinlan Terry chocolate-box mock Georgian. So he designed that and we got planning permission, and straight away I decided it was horrid.

So we redesigned the entire frontage, got planning permission for that too and then decided it was slightly in the wrong place. So we got planning permission to move it six feet to the west, and then, very nearly two years ago, work began.

Covid and the first lockdown meant the original completion date of May 2021 was pushed back to July. And then something else – I'm not sure what – pushed that back to the end of August. And then I was given a date of mid-October, and now I'm hearing rumours that I won't be in till the standard builders' finishing time of 'Christmas'.

This would be acceptable if they were just getting on with it. But this is the thing you don't realise when you're constructing a house: builders are like six-year-old children. They need constant supervision. If you let them 'get on with it', they fall in ponds and spill paint and drop stuff, and then they all argue and sulk and you have to stop what you're doing to sort them out.

Sadly you will not have time to do this because, to have the millions of pounds needed to build a house, you will need eight jobs. That's why I write three newspaper columns a week, I do a car show, I do a farming show, I run the farm – not that this pays for much – I have an internet business and I present *Who Wants to Be a Millionaire?*.

But still I'm expected to spend two or three hours a day dealing with the builders. Who talk in a language I don't understand, peppering their sentences with acronyms and technical terms. And they all use a weird Roman Catholic system of measurement, which, because I think in feet and inches, means my front door is taller than the Eiffel Tower.

James, my excellent project manager, says that the builders I'm using are actually pretty good, even though they forgot to order a pantry door, the staircase arrived and was six inches too short and the cast-iron spindles looked as if I'd made them in metalwork at school in 1974. Oh, and the spare bedroom door is in the wrong place, the brilliant Romanian chap went home, the wall in the hall wasn't straight, the stonemasons fell out with the contractors, the site manager walked, the ashlar cracked in a winter frost, the garden wall fell over 24 hours after it was built and the bathroom floor is still a tree.

Two years ago James looked like Harry Potter. Now it's as though Michael Foot is walking through the door every morning. But for me it's worse, because even when things are going smoothly – and they did, back in

March, for seven minutes – I have to be on constant call to make decisions.

I've written here before about the enormous amount of time and effort I put into choosing the door handles: the precise colour and texture, the feel, the grip and whether to go for bronze or steel. I put in a similar amount of time picking out the lavatory seats and the light switches. But that was two years ago. Now my standard answer to everything is: 'I don't care. Just build the effing thing.'

I've been living these past three years in what is officially the smallest cottage in Britain. I can stand in the centre of the sitting room and touch all four walls. And I'm able to cook supper and go to the loo simultaneously. It was awful to begin with but now it's intolerable.

And what makes it all so much worse is that progress on the new house has slowed from a crawl to what a tectonic plate would call 'pedestrian'. I went to Capri for a few days last week, and when I got back I expected them to be oiling the letterbox hinge and tuning the television.

But it was as though they'd spent the entire week in the playground. To my untrained, feet-and-inches mind, they hadn't done anything at all.

The problem is that the act of actually building a house is pretty quick. Putting one stone on top of another is the work of a moment, and putting the roof on – even if the stone tiles are 200 years old – is incredibly speedy.

But now we are at the stage when they are setting up lighting boards and shower attachments, which is fiddly and invisible and so slow that by the time they're finished they'll need to install a Stannah stairlift so that I can go to bed at night.

We are often told that building your own house is one of the most satisfying and warm, fuzzy things you can do.

But I can assure you, it's complicated and slow. You are much better off buying one that was built many years ago, by a Georgian.

My cunning plan to help tackle climate change

Is anyone else mildly amused that in our drive to reduce the amount of carbon dioxide we produce, we've ended up with a shortage of carbon dioxide? This means food cannot be frozen or cooled as it's moved to the supermarket. It means animals cannot be stunned before they're executed. It means Pfizer vaccines can't be stored, surgeons can't operate on your liver, fire extinguishers can't be refilled and fizzy drinks can't be made to fizz.

Ostensibly, all of this has been caused by an American outfit called CF Industries, which, thanks to rising gas prices, has temporarily closed two UK fertiliser plants where CO_2 and dry ice is manufactured. But actually it's yet another example of why our drive for green energy can only work if the world order remains untroubled by events. Which is highly unlikely.

As I write, our much vaunted windmills aren't turning because it isn't very windy, and we can't rely on coal-fired power stations because they're all being closed down. And we haven't been able to build any new nuclear reactors because of some newts. Which makes us reliant on gas.

And that's a problem because a burly Russian with a beef about something or other is standing on the hose that delivers natural gas from the Urals to Europe. And to make matters worse for our energy needs, the cable that brings electricity from France to Britain was damaged by a fire, and it won't be mended for the best part of a year, partly because Britain can't mend anything these days – see Hammersmith Bridge – and partly because the French want to pay us back for the submarine deal we did with, er, that fella from down under.

So here we are, facing a cold and lonely Christmas, eating whatever mouldy pickles we can find at the back of the larder, and not being able to visit our relatives because we can't afford the electricity needed to charge our electric car.

To sum up then, it's very easy to say, 'We shall be carbon neutral by 2050,' but it's very hard to do because you can't ever see what problems are around the corner. And nor can anyone ever predict the consequences. I mean, who could have guessed when those Russians smeared bits of novichok all over Salisbury that, a couple of years later, we'd be unable to fill our fire extinguishers? Or that we'd have to drink flat cola. Or that surgeons wouldn't be able to stabilise our meaty centres when doing invasive surgery.

And I'm afraid it's going to get much, much worse because Carrie Johnson and her bouncing Boris mouthpiece are about to turn their eco-guns on farming.

They've established that Britain's farmers produce 10 per cent of the nation's greenhouse gases and have decided that it would be better if these ruddy-faced planet killers were ethnically cleansed from the countryside. Then it could all be sold to Russian oligarchs, who'll plant a billion conifers, and eco-lunatics, who'll fill the resultant monoculture with wolves and bears.

Couple this to the ban on internal combustion and the exciting plans to make everyone heat their homes using soil and, yes, it's possible that Britain will be carbon neutral, as promised, by 2050. Lovely. But where will our food come from?

Today, Britain is 60 per cent self-sufficient and there are calls for the government to not let it fall below that figure. Calls that, I bet, will be ignored because making food using current technology produces global warming. It comes from the soil when it's turned and from the tractors that do the turning and from sheeps' bottoms and cows' mouths. And I'm sorry but Carrie has made it plain that achieving carbon neutrality by 2050 is not just a government priority. It is *the* government priority.

So, right, we buy our food from abroad. Which means we've simply exported the global warming problem to another country. That's the nuisance fact about climate change that I don't think Carrie has grasped. It isn't going to be fixed if a small rock in the North Atlantic goes all happy-clappy. It has to be a planetary thing.

And there's more. We discovered at the beginning of the coronavirus pandemic that the world is now so interconnected and interdependent that someone eating a bat in Wuhan will cause Britain to run out of lavatory paper. We are discovering now that a small fire in an electrical substation can cause us to run out of fizzy drinks. And who knows what we will discover down the line if we decide to exist on imported food?

There was a school of thought in the thirties that Britain should concentrate on making stuff out of coal and steel and steam and simply buy food from abroad. But then, out of nowhere, the U-boats came and we all damn nearly starved to death.

It's reckoned that if there was an interruption to food supplies today, we'd have three days before the looting started. Soon after that Robert Carlyle would be trying to eat your arm.

It's strange. We run a Ministry of Defence to ensure British subjects are safe. We have aircraft carriers and nuclear-powered submarines and fighter jets and tanks and atomic weapons and thousands of soldiers. We spend billions every year on this even though the chances of us actually being invaded these days are extremely small.

But the chances of there being an interruption to our food supplies are actually quite high. It could be one of a million things. The workers who normally pick our fruit and vegetables are stuck in Lithuania because some old people in the north don't like living next to a family from Pakistan. There is a shortage of lorry drivers. Fertiliser factories are closed because the Russians are being petulant. There's a container shortage. All the ships are stuck in the Suez Canal. There's a pandemic. You put two or three of those things together and don't think you'll be able to go to the Winchester for a nice cold pint until it all blows over because there won't be any beer.

So if our food supply is precarious, and it is, why is the government not hurling money at the issue, the same way it hurls money at the MoD? I suspect it's because it doesn't know how that money should be spent.

Happily, I have an idea. The government is taking away the single farm payment, which means farmers will no longer get money simply for owning land, and says that in future there will be 'public money for public good'. They haven't actually explained what this snappy marketing slogan means, but we all sort of know already:

they're going to give us cash if we rear baby hedgehogs and offer free sunflower seeds to passing ramblers.

In my mind, though, 'public good' means 'public security'. The knowledge that if the shit hits the fan, for whatever reason, this country can feed itself.

Of course farming must play its part in the war on climate change, but this will require huge investment in new technology. Robot tractors that can kill weeds with electric shocks rather than chemicals. Tech that allows crops to grow without soil. And a whole new supply chain that minimises the miles food travels.

This, then, would be a sensible thing to do. Replace the old-fashioned farming subsidy idea with grants that encourage farmers to come out of the red phone box and into the iPhone world. A world where farming is more efficient, more environmentally friendly and which, no matter what happens, is capable of feeding all the people who live here.

Pheasant plucker

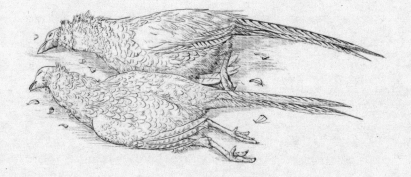

As we are now in the season, it's a good time to address the issue of game shooting. I enjoy it very much but I'm well aware that all of the nation's vegetablists – and several enthusiastic meat eaters as well – think it's disgusting that, in this day and age, a pack of usually drunk Hoorays are allowed to rampage around the countryside in tweed shorts and Range Rovers, killing God's little creatures for fun.

I respect this point of view. Because if I reared puppies on my farm and then released them from their pens so my mates could try to shoot them as they ran for cover, I would be sent to prison, and rightly so. But the pheasants and partridges I do rear, and then release from their pens so that they can be shot, are not dogs. They barely qualify as creatures as they have the intelligence of an ironing board and the personality of a Liberal Democrat.

There's more. If you are a carnivore you will accept that an animal should be dead before you break out the horseradish and brandish the knife and fork. Which means someone has to kill it. So what are you saying? That the person charged with this task must be unhappy about it? That he or she must share no jokes with their workmates at the abattoir? And that there should be no gentle flirting round the water cooler?

Of course not. So if you are saying that a person can be happy as he goes about the business of turning a cow into a beef, then why can I not have fun while shooting my pheasants? Or to be precise, shooting at the bit of sky where, moments earlier, a pheasant had been. I'm not a good shot.

Although there are worse. There's a story I once heard about a famous jockey who went on a pheasant shoot. I won't say his name here, save to say it begins with a P and ends with an iggott. Anyway, when a bird walked out of the woods in front of him, he raised his gun and began to track it. His loader, standing next to him, thought he was doing this for a laugh and only began to worry as the bird approached the line of 'guns' – they're the people in the tweed shorts – because the one thing you never do, apart from shooting a bird on the ground, is point your gun at someone. Ever.

Eventually, as the gun swung towards the people, the ruddy-faced instructor was forced to intervene, asking what our nameless friend was doing exactly. And he got the reply: 'I'm waiting for it to stand still.'

That's the thing about pheasant shooting. It's not supposed to be easy. The bird must be flying when you shoot at it, and even then not just cruising around at head height. Smoking a low hen is considered very poor form. It must be high, like Telstar, and flitting in and out of the clouds as it screams towards you doing 45mph.

This means you must not aim directly at it, because by the time you've decided to fire and you've pulled the trigger and the shot has covered the distance to the bird, it'll be long gone. You need to shoot in front of it. A long way in front. I've been told that in the war anti-aircraft gunners had to shoot a mile in front of a Heinkel if they wanted to hit it. A mile!

Working out how far in front of a bird you need to be is tricky. Because you have to calculate the bird's speed and the wind and consider the fact that over a distance of 70 yards your shot will droop by maybe 14 inches. And you've got to do that in a millionth of a second. Incredibly some people can, even with low-powered 20-bore shotguns. Some can even work out trajectory, knowing precisely where to shoot the bird so that it lands on a mate's head. AA Gill used to try and do this to me all the time.

Luckily he was also a useless shot. Because a pheasant crashing into your head at 45mph will kill you. A bird I once shot ended up on a chap's Range Rover and afterwards it looked like someone had crash-landed a helicopter gunship on the bonnet. The damage was huge.

And now you're feeling uncomfortable because you're reading about the slaughter of birds in the name of japery. Hmm. Yes, there are shoots where the birds are bulldozed into the ground after the day's drinking is over and that's indefensible. But on my shoot all of the guests and all of the beaters go home with their supper. We really are shooting food. Except when I once got bored and shot a trout. That was pretty much inedible.

And it's not food that was reared in a shed, in artificial light, up to its knees in its own faeces like the chicken you're having for lunch today. A pheasant can fly away at any time, but it chooses to hang around because it's fed and watered and given a home. It has a genuinely happy life.

Is there a better way of killing it than shooting it? Well, I suppose I could sneak up on it in a camo suit and strangle it, or beat it over the head with a stick, but would that be better? I'm not sure it would. Given the choice I'd definitely prefer to be shot.

And there's more. When someone has a shoot, they manage the woodland in which the birds live more carefully than if they did not. They clear away what's harmful and leave wild patches around the edges and all of this makes life better for insects and other birds too.

You watch what flies out of a game crop before the pheasants become airborne. Hundreds and hundreds of songbirds, birds that would have nothing to eat and nowhere to shelter were it not for Rupert, Rupert, Rupert and Nigel.

And from this year onwards the meat shoots produce won't be so full of lead because lead shot is being phased out. We must use alternatives such as steel instead. Not sure why, as I can't imagine the birds care either way. Crows do. They know exactly what gun you have and what shot you're using and always fly just out of range. Generations of pheasants, on the other hand, keep coming at you, hoping that if they do the same thing over and over again the result might one day be different. Which it is if they fly over me, because I usually miss.

I get that Chris Packham wants pheasant shooting banned but I think he's wrong. I think that if there were no shooting, landowners would be less bothered about looking after their land. I think a great many countrymen who earn a living on shoots would lose their jobs.

I think woods would be bulldozed to make way for something profitable, and I think we'd not only have less choice of what to eat, but also that what we were offered would somehow be less wholesome.

So, sure, campaign to end shooting by all means, but know this. What you're actually doing is waging a class war. You're not trying to make a pheasant's life better, because it's already very good and frankly you don't really care either way. No. What you're actually doing is trying to make a Rupert's life worse and that, to me, seems a bit petty . . .

How to crash a Lamborghini tractor

My first crash wasn't so bad. I was heading from the Tan Hill Inn in North Yorkshire to a village called Keld and then all of a sudden I wasn't any more. Instead I was bumping over a grassy knoll and then some rocks, while listening to the front wheels on my mum's Audi coming off.

After the crash I quickly established that my passenger was not hurt and then, having climbed out, that the car was. Certainly it was no longer driveable, and this is when things got bad. Because I knew at this point that I'd have to fess up. My mum didn't know a lot about cars, but she was bound to notice that the front wheels on hers were missing. And then she'd be cross and point out that I'd only passed my driving test 36 hours earlier and was an idiot.

It's interesting. In earth years the gap between having a car accident and then telling your mum may only be 30 minutes but it will feel like four trillion years. Entire geological periods pass more quickly. It's enough time to witness the birth, rise and subsequent extinction of an entire species.

But it's not as long as the gap between crashing your tractor and sitting there waiting for Kaleb to arrive on the scene. Because he was bound to ask how, in the name of all that's holy, I'd managed to hit a lone telegraph pole in a 52-acre field.

I had to admit that it would be a good question. And I also had to admit that I didn't really have an explanation. It had been foggy, yes, but it wasn't like being immersed in a bag of oxtail soup. I saw the pole coming from way back. And I watched it grow larger in the windscreen. And then I hit it.

Actually, that's not strictly accurate. I didn't hit it. Nor did the tractor. But a wing of the six-metre-wide disc the tractor was towing did, at 17kph. The damage was substantial. Not only to the pole – which carries a 13,000-volt cable, by the way – but also to the disc. Two of the box section steel arms were pretty much snapped off. And it wasn't mine. I'd borrowed it, which meant I'd have to ring its owner, who's the uncle of a friend, and admit that I'd written off his £20,000 machine. Although when I say 'I'd have to ring him' what I mean is 'I'd have to ask Cheerful Charlie, my land agent, to ring him'.

He'd also have to call Scottish and Southern Electricity Networks (SSEN) to say that a 13-tonne rig had just bashed into one of their poles. But before that I had to wait for Kaleb.

I saw the cormorant die out and a new species of winged platypus come and go. I saw the next Ice Age and the intense volcanic activity that followed. And then finally a pick-up truck came bumping over the field containing a man with white knuckles and a tempersome face.

'How've you done that?' he asked crossly, before the door was shut.

'Dunno sir,' I mumbled.

'Can't you do anything properly?' he wanted to know in a hands-on-hips type of way.

He did have a point on that one. Last year I swung off the farm drive on to the road and had travelled several hundred yards before I realised that I was towing an entire hedge in my wake. And then there was the barn I'd bent while trying to reverse a trailer into it. Oh, and a few bins I'd knocked over, and some gateposts.

It's odd. I'm not a particularly clumsy driver when I'm in a car, but in a tractor I almost never get a day's work done without hitting something.

Mostly, I suspect, this is because tractors are surprisingly hard to drive. This is because they don't drive over things so much as bounce, which means that when you're behind the wheel you're having a constant battle with gravity. One minute you're smashing into your seat like you weigh 30 stone and the next you're floating around like Jim Lovell in Apollo 13.

Only, unlike Jim Lovell, you're in a wood, surrounded by trees and you don't have Velcro-soled boots to keep you anchored to the floor. Plus, all he had to do was get the re-entry angle right by lining up an entire planet in the window, whereas in a tractor like my Lamborghini, which has two brake pedals, four gearlevers, two throttles and 48 gears, you are constantly having to adjust something while not bashing into a nearby wall.

And that's hard in my tractor, partly because its brakes don't really work, no matter which of the pedals you press, and partly because for the past few weeks I've been discing the fields to keep Greta Thunberg happy. In simple terms, this is eco-ploughing. Only the disc I have weighs more than four tonnes, and it's mounted rather than towed, which means the front wheels of the tractor are mostly airborne. So I don't have much in the way of steering either.

Controlling this bouncing nine-tonne monster, then, is a pretty hard job, and that's before we get to the business of operating whatever machine is attached to the back. Which, because there are so many buttons, is like playing motorised Pelmanism while on board a crashing airliner. And don't even think about pressing the wrong button, while smashing into your seat at 40mph, having moments earlier bashed your head on the glass sunroof, because that will turn on something that will cause a scar in the field that'll take a whole year to mend.

When you come up behind a tractor on the road and you wonder why the chap in the cockpit doesn't pull over to let you by, it's because he's operating at the very limits of what a human can do. Well, I am, at any rate.

And that's what I told Kaleb, who didn't seem to be listening. Instead he was looking around, trying to work something out. And then, after a period of sweaty silence, he said: 'Why were you discing this bit of the field anyway? I did it last week.'

I hadn't even had a chance to answer before his voice went up a couple of octaves and he really let rip. There was something about 'you've been doing this two years now' and 'you should be getting better' and after a bit of swearing and name-calling, there was an observation that even a child can tell a bit of field that's been disced to a bit that hadn't because 'one bit's brown and one bit's green'. And this was all topped off with 'you're banned from tractor driving from now on'.

With that he stomped back to his truck, which I then noticed for the first time was white. Which is odd because his normal ride is black.

Later in the day – after SSEN had said the telegraph

pole didn't need to be replaced – I made some investigations, which revealed that his truck was at the menders because while coming down his drive that morning he'd been checking his phone. And had driven into a wall.

A house move from hell

Even by my own dizzying standards, last week was busy. Due to the fact I hadn't read the small print in my contracts, I was legally obliged to whizz about on a promotional tour for my new book and for the next instalment of *The Grand Tour*.

And on top of the cheery chats with Jonathan Ross, Chris Moyles, Zoe Ball, Steve Wright and everyone else with access to a microphone, my cows needed impregnating, and I had a day's shooting with my neighbour and a book signing at my local bookshop. Plus the puppies needed their second vaccination, my car needed servicing, I had to do a photoshoot for the *Sunday Times Magazine* and there were three newspaper columns to write. Oh, and on Wednesday I was moving house.

It's claimed by the weak and the unbusy that a house move is the most stressful thing you can do – apart from dying in a car crash, presumably – but I reckoned I had all the bases covered.

As the builders were supposed to have finished at the

end of July, I'd had plenty of time to do all the necessary shopping. I'd done a thorough trawl of Peter Jones to buy all the kitchen and bedroomy things I'd need, and in addition I'd been to various house sales and to a place called Lorfords, in Tetbury, which is two gigantic hangars full of a billion things I wanted. Including an 8ft-long model of a French railway station.

I'd even spent some time with a local and quite brilliant taxidermist, from whom I'd bought a fireguard full of brilliantly coloured hummingbirds, and a stuffed lapwing. And over the past few months everything was delivered and neatly stored in various barns waiting for the Big Move Day.

Obviously, as the house was being built by builders, the wait was long and littered with broken promises. But eventually they said they'd be finished in mid-October, and then, when they missed that deadline by a mile, they assured me that if I gave them just one more month, they'd definitely be done.

The month was up on Wednesday, so the moving men turned up to find that the staircase banister wasn't fitted, the wallpaper man had gone on holiday and both the joiner and the electricians had gone Awol. Not since Schubert has the world seen anything less finished. But, as someone once said, the only way to get builders out of your house is to move in. So that's what we did.

It was harder than I imagined, because everything coming out of the cottage where we've been living for the past four years was more disgusting than a diseased sheep's rear end. We found a putrefied rat in one of the armchairs, six inches of dust on everything else and, at the back of the kitchen cupboard, a Yorkshire pudding baking tray that had been painstakingly converted by a mouse into a two-bedroom house. He'd shredded a takeaway menu to make a bed in one section and had used another as his lavatory. He'd even turned a pan lid into a roof so Lisa's cat couldn't kill him.

Bit by bit, though, everything was dusted and cleaned and placed in a box by the endlessly patient moving men, and it was like being reunited with my whole life. I found an enamel nameplate saying 'Jeremy's room', which used to hang on my bedroom door when I was little. I also found my school boater, the laser pen I used to annoy Lisa's useless cat and a lot of fondly remembered clothes that must have shrunk somehow while they were down the back of the shoe rack.

Soon everything was in the new house, and then the laborious process of deciding where it all should go began. A process not helped by having to break off every 15 minutes to have Zoom calls about the farm show or to hold the puppies while they were injected or to deal with some halfwit with a long lens taking pictures for the endlessly unpleasant *Mail Online*. And then rushing back to find the photographer from the *Sunday Times* wanted me to wear a Father Christmas outfit. Top tip: when you're moving house, clear your diary for that day.

Eventually everything was in the right room, and the even more laborious process of unpacking began. We began with the boxes that had been delivered from Peter Jones. In the first there were four sets of knives and forks and, in the second, four more. In the next three there were more sets, until eventually the cutlery drawer was full.

We then found several more sets, and at this point I was to be heard from several miles away wondering loudly why we'd bought so many. And what we'd been thinking of. And why Lisa hadn't stopped me, for crying out loud. And Lisa was standing there with her hands on her hips explaining that we had started shopping at Peter Jones after a particularly long lunch at the Colbert café.

This would also explain why we'd ended up with nine beds. Which is more than you need in a six-bedroom house.

The worst excessive buying was of tumblers. Somehow we'd bought 70, and each came with a sticker that was nigh-on impossible to remove. That's all I did for one night. Peel off the stickers using nothing but swearing and then put the glasses in a cupboard. And all Lisa did was take them out of the cupboard I'd chosen and put them in another one.

We had a row about that, and then we had another about why the corkscrew had been packed in a random box and not kept separate, because we were definitely going to need it before anything else.

Today, apart from writing this, what I've done mostly is wonder what to do with the 8ft railway station.

But I did take a moment to stand in the middle of the kitchen, thinking that, while moving house has caused several arguments and some bruised fingers and is going to take months to finish, it's worth it. Because you're moving into a new house and you're going to be very happy there.

The great escape

All my veals escaped last week. Actually, that's not strictly accurate. A veal is a young cow that's dead and on your plate, all covered in onions and jus. Whereas my young cows were not dead when they escaped so they were technically calves. I think.

I still haven't quite got my head round all the bovine terminology. I thought you got dairy cows that produced milk, and beefs that produced meat. But that's like going to a sixth form debating society and saying that there are only two kinds of human: men and women.

In the cow world there are stores and heifers and bullocks and steers and bulls and calves and, to make matters even more complicated, whenever one of the group is menstruating (can I say that?) all the others immediately become LGBT and spend all day riding one another round the field like wheelbarrows.

I even have one youngster who is transitioning from male to something else. I'm not sure what exactly, but the suggestion from experts is that its testicles, which were supposed to have been crushed months ago, are beginning to emerge once more and that unless I have a fact-finding rummage round its rear end very soon, it will probably impregnate one of its sisters.

First, though, I had to find it. I had to find all of them, in fact. Because in the field where I'd left them there was nothing at all apart from a big hole in the fence. The whole flock had gone. All nine of them.

It seems that cows, like dogs, enjoy being tickled, but as this is not possible when you have hooves, what they do is rub their flanks on fences until the fence gives way.

Happily, this gave me the point at which they'd got out and as the ground was pretty soggy I was able to track them into a nearby wood. They'd tried, like Harrison Ford in *The Fugitive* and Tim Robbins in *The Shawshank Redemption*, to throw me off their scent by walking down a stream, but at some point they'd had to get back on the bank and immediately I was on to them once more.

By examining broken branches and snapped shrubs I was able to follow their path for a good 200 yards, until I got to a badger sett, where the footprints went into the hole. Deducing that I had been following the trail of the wrong animal, I made a note of the sett's location – for no reason at all, d'you hear – and went back to resume my search at the stream.

Then I had a call from a neighbouring farmer to say that he'd found them in one of his fields and, as a former dairyman, he'd been able to herd them into his yard, where they were waiting for collection. 'Have you got a trailer big enough for all eight?' he asked.

Eight? Yup, one was still missing. And suddenly my tracking job became much more difficult. I know that Lord Baltimore in the Butch Cassidy movie was able to determine, just by its hoofprints, whether a horse was being ridden or not, but I hadn't been able to tell the difference between a cow's footprint and a badger's. So I definitely wouldn't be able to work out when the solo operator had broken away from the rest of the pack.

Still, you're thinking, even a young cow is pretty big, and you'd be right. But Oxfordshire is even bigger. And a cow is like a deer or a sniper: it's able to blend in. You could be right on top of it and simply not know.

I thought about using my drone, but this had been converted into a barking sheep herding device two years earlier and then thrown into a bin. I called my other neighbours to see if they could help with the search. But they all pretended to be out. So I couldn't use their helicopters.

I didn't even have a film crew that day, so it was going to be me and Lisa and a local whom we shall call Donald. Because that's his name. And after an hour, while Lisa and I were still in the wood, examining every footprint to see if we could establish a direction of travel, Donald called to say he'd found the veal about two miles away.

We hurried over in the Range Rover and arrived just in time to witness quite the most bizarre car chase in all of automotive history. Donald was on his quad bikey Mule thing, with the engine at full chat, chasing the cow, which was going so fast it had vapour trails coming off its ears.

Did you know cows can travel at 70mph? Neither did I. And nor did I know that they can jump. Not just a little bit but clean over a 5ft wall. I swear to God that if you entered a cow in the Grand National it'd win by half a mile.

After half an hour of this high-speed pursuit, during which time we overtook the London Paddington to Hereford express, we were nearing the Devon-Cornwall border, but finally the animal began to tire and was happy to be walked back to join its mates in the farmer's yard. And now it's back in the field, scratching the fence with its arse.

And stubbornly refusing to let me examine its nut-sack. I've bought it a scratchy thing, which is a series of wire brushes on a pole, and a delicious mineral lick, but it's all to no avail. It's the only straight male in the flock and won't let me anywhere near it. I shall have to wait till its sister has a period and then it'll become gay.

Meanwhile, my lady cows are somehow not producing milk. I have no idea why this is so, but I'm glad because I have no means of collecting it.

And I thought sheep farming was difficult and complicated. Sure, cows don't actively want to die and you can go to a party knowing that there's very little chance of getting a call after the prawn cocktail to say that one of them has got its head stuck in a terrier hole. A terrier's hole maybe. But not the hole a terrier has dug in a bank.

That said, though, they do require thought and attention and there's the constant worry that they'll get on a road or, worse, they'll catch bovine TB. A problem that can't be solved very easily when you're being filmed.

Still, at some point next year some of them – I'm not sure which – will go to the abattoir, where they will be converted into delicious cuts of meat that I can sell at my new restaurant. If I'm allowed to build it.

Because while Lisa and I were busy rounding up the cows, like Pete Duel and Ben Murphy, 28 of the local villagers were sending their objections to the council. Let's hope the council realises that this means 800 people in the village didn't object, because they fancy some well-bred, well-reared, grass-fed local beef. And they quite like having someone nearby to look after the countryside.

Grow your own beer

As farming subsidies are very firmly on the way out, the nation's tractorists must find new ways of keeping their heads above water. Which is why I've decided to turn a large chunk of my land into beer. Yup. I've invested in a brewery.

I should explain that I'm not making what I like to call a 'James May beer'. It's not something that the Campaign for Real Ale would support, because it's not brown and it does not have twigs and mud in it. Drinking stuff like that is like drinking a meat pie. So my beer is a lager.

Choosing a name for it was the first problem. I wanted to call it Lager McLagerface. Or McFace for short. But one of my partners in the enterprise is an important London ad man, who said that McFace didn't conjure up quite the right premium image. He wanted the Milk Tray chocolates man parachuting into a schloss, not Begbie and a mega mega white thing.

Eventually we settled on Hawkstone, because there's a neolithic standing stone of that name here in the Cotswolds. And it was the neolithics, 4,000 years ago, who invented barley farming. Weirdly, back then, they just ate it.

They didn't realise it could be converted into an alcoholic beverage, which is why, I suppose, they spent their evenings upending rocks rather than getting pissed. These are the guys we have to thank on a summer's day when we're in the garden and all we want is a glass of cold, refreshing lager.

But then, while we were celebrating our genius with the name, Cheerful Charlie, the land agent, turned up with the crop plan he'd devised for my farm this year. There were three types of wheat, two types of oilseed rape, echium, potatoes, grassland for the cows and some winter barley. But no spring barley at all. I don't know why beer has to be made from barley that's planted in March rather than September, but it does. And Charlie had left it out.

'You've left out the spring barley,' I said.

'I know,' he replied. 'It's always disastrous on your farm, so I didn't think we'd bother any more.'

At this point I pulled my special crestfallen face and explained that I'd just invested in a brewery that would use my barley and that we had a name and everything. 'Ah,' said Charlie, before gently rolling his eyes and rhythmically banging his forehead on the table.

It turns out that spring barley is fickle. It gets stroppy if the weather's too cold or too hot or too wet or too dry.

And it really doesn't like the brashy soil on my farm, so growing it is a risk. It's also susceptible like a sickly child to all sorts of disease, and if there's too much nitrogen in the finished product, it will be rejected by the grain merchant and will have to be used for animal feed. So you work hard all summer nurturing the bloody stuff and then you end up selling it for 3p.

Last year I was lucky. It didn't grow to be especially tall, and there were plenty of bare patches in the fields where I'd planted it. But I got a good price of £205 a tonne. Or at least I thought it was a good price until I discovered that the brewery is paying £580 a tonne.

So I take all the risks, I do all the hard work and I get less than half what the maltster is paid for wetting the grain a bit to make it germinate and then heating it up to make it dead again. So many farmers round here are so outraged by this that we are having a meeting to discuss setting up a co-operative malting operation. Shouldn't be difficult to find a premises. Every village has a house called The Maltings in it.

But that's then and this is now, and everyone at Diddly Squat is fully engaged with the idea of diverting all our energy next year into careful spring barley farming. Even Charlie, who when I talked about it this morning said: 'Oh God.'

The next big issue was the recipe for the lager. It turns

out that beer has only four ingredients – malted barley, hops, yeast and water – but it's very easy to make a complete pig's ear of it. See Budweiser for details.

The first job was to work out what alcohol content we'd like. I consulted my 25-year-old son, who said that he favoured lots as he doesn't have much money and wants to get as pissed as possible for the smallest possible outlay. Then I talked to my dry-stone waller and head of security, Gerald, who said that when he goes to the pub he likes to have a bath first and put on his smart clothes and it's not worth doing that if he's only going to drink a pint. He likes to have a good old natter with his mates and wants a gallon. So he wants the alcohol level to be as low as possible. 'Grrr erdickle sisanatious,' he said.

Interesting dilemma. Who did we want as a customer? Gerald or my son? We decided on both and therefore settled on an ABV of 4.8 per cent. That sounds like a nice number to me. You'd certainly want a car with a 4.8-litre V8.

Luckily, Rick and Emma, the people who run the brewery, can handle requests like this. Rick is the boffin. You can tell this because he wears white wellingtons and has mad hair that I think he cuts himself. His wife, Emma, runs the business. And now they have me on board as well, so things should really take off.

The start wasn't smooth. We had a blind tasting in the

summer and everyone agreed that our new lager was nowhere near as nice as the beer Rick and Emma had been making for years. So Rick went back into his room full of pipes and tanks and started again, and I'm not just saying this: now, it's really good.

So we had the beer and we had the name and we had a farm to produce the barley and we had plans to cut costs by axing the middleman. We'd even worked out that we could sell it in Lisa's shop and on Jeff's website. The ad man even reckoned he knew a few people in London who might stock it.

So all we needed was a television advert, which I said I'd do. I booked the film crew, called the farmer who owns the land where the Hawk Stone stands and wrote the script. And this morning I received news from some lawyers pointing out that you're not allowed to say in an ad that 'Hawkstone is just what you need before a hard day at work'. And neither can you take a swig and say: 'F*** me, that's good.'

This, then, is the problem farmers face. They are told by government that to survive they must diversify. But when they try to branch out by turning their own barley into lager, they are forced by law to pull a po face and say it must be drunk responsibly.

WINTER

Christmas at Diddly Squat

Christmas on the farm. You can picture it, can't you? A steaming home-grown goose, glistening in the candle-light. Piles of spuds from the garden and buttered vegetables from the fields. Rosy-faced children playing happily with the wooden toys you whittled from trees in the forest and, after a sticky toffee pudding, some hearty toboggan rides in the snow until the watery sun's chilly demise sends everyone running inside for some fireside sherry and the shoulder-punching, yuletide jumper fun of having to mime *The Beastly Beatitudes of Balthazar B* in a game of charades. Idyllic. Nearly as idyllic in fact as having to act out the trickiest charade of all time, *Versailles: The View from Sweden*.

Sadly Christmas on my farm won't be like that. It will be like every other day, only with added mud. When I first became a horny-handed son of the soil two years ago, I figured that winter would be an easy season. The sheeps would be pregnant, the cows would be in their house, the crops would be growing on their own, the hens would be standing around waiting to be eaten by a fox, the badgers would be spraying TB around the fields and I'd be in Val d'Isère, living it large on the fat subsidy cheque.

That's not how it turned out, though, partly because the subsidy cheque is now smaller than a prewar postal order and soon it'll be even smaller than that. And also because winter is actually the season when you go out into the cold and the rain and the mud and do all the unpleasant jobs you should have done in August while you were waiting for the crops to dry.

Mostly this involves looking at broken fences and gates, hoping that somehow they will mend themselves. They can't, so I have to do it and, amazingly, this is a skill I don't have. While I've always said that hammers can be used to fix everything up to but not including a child's poorly eye, I've never really understood how they may be used to drive a nail into a piece of wood.

On the first blow I always catch the side of the nail-head, causing the nail itself to bend. And then I imagine that if I strike the opposite side of the head with the same level of ferocity, the kink will straighten out. It never does though. It gets worse and worse until eventually I'm forced to hammer it in sideways before reaching for another nail, which does the exact same thing.

Soon I will forget to concentrate and bring the end of the hammer down on my thumb, but that's OK because when you are on a farm in the winter you are so cold you don't even notice when you've just smashed one of your own bones. I've heard about some farmers cutting off their right arms and not noticing until they get in the tractor to go home and find they can't engage a gear.

Eventually, though, after you've smashed most of the bones in your hand and bent two whole bags of nails, you will get one to go through the piece of fencing post and then you'll hold up another so that they can be joined for ever more by the nail. Nope. What actually happens is the second bit of wood splits and then someone rural strolls by and says, helpfully, 'That nail's too big.'

Last year, while attempting to nail a piece of fence rail to a post that was already there, I was so fierce with the hammer, I knocked the post clean out of the ground. And honestly I felt like sinking to my knees and weeping. It's just so frustrating when you are hopeless at whatever it is you've chosen to do. It really is.

Of course you would imagine that on Christmas Day it's possible to take a day off fence and gate repair work, but I'm afraid not. Because the birth of the baby Jesus is of no concern to the sheep, who want to escape and die every day of the year, or the cows, who just like knocking fences down.

Back in November I came back from a weekend of

solid drinking and staying up late in Scotland, absolutely knackered. I fell into bed on Sunday evening at around 10.30 and was woken up at 2am by an alarm on my phone, alerting me to the fact that the automatic hen house doors were opening. It took an hour to sort that out and I didn't get back to sleep till four. Then, two hours later, I was woken again by a neighbour who said my cows had pushed over the fence again and were on the A361 two miles away. And there's nothing on God's green earth I can do to be certain that the exact same thing won't happen again on Christmas Day.

Look at *Jurassic Park*. To make sure the dinosaurs stayed in their pens, Dicky Attenborough built a trillion security systems into the perimeter fences, all of which also had important-looking yellow flashing lights on them. And we all know what happened next. Martin Ferrero went to the lavatory and got eaten.

To stop that kind of thing happening at Diddly Squat, I've already bought my cows lots of distracting presents. There's an £800 scratchy thing that looks like an upended carwash roller, which they can use instead of fence posts to reduce itchiness, and two footballs that I fill every morning with nuts and silage. Nothing works because animals don't understand presents either. They look at their new stuff and then walk through the fence again. And it'll be the same story at Christmas, I just know it.

Even if I buy them some new trousers and the latest *Call of Duty* video game, they'll be off and away into the village to eat all the local Hamishes' Brussels sprouts.

And even if they don't escape, they'll need hay and the pigs will need potatoes and the hens will need their grain and the pheasants will need moving and the new puppies will need walking and the trouts will need feeding and the sheeps will need demaggoting, and when I've done all that, in the cold and the rain, I'll be so tired I'll fall asleep with my face in whatever Lisa found that morning in the freezer. Mutton usually, from one of our sheeps. Which gives me heartburn. Yup, even from beyond the grave those bastards continue to make my life miserable.

Still, could be worse – my Christmas could be like yours. Because you'll get back from the pub on Christmas Eve at quarter to two, and fifteen minutes later one of the kids will be sick on you because he has eaten all the sweets Santa brought, and then after a morning of arguments during which all the other presents will be wilfully destroyed in a series of bitter sibling arguments, you will open your presents to discover they're terrible before cooking half a hundredweight of food the kids don't like and then trying to digest it while watching a kiddy-friendly film about an otter with a spade in the back of its head.

Jeremy's Christmas list

- **Will give Lisa** – a sculpture. Last year I scored a hit buying her a Nic Fiddian-Green horse-head piece.
- **Will give Kaleb** – I know he wants a Fendt tractor so I'm getting him one – a model one made by Britains.
- **Will give Gerald** – He wants a gas-operated parsnip thingummy but they don't have them at StowAg.
- **Hoping to receive** – Ken Miles's Ford GT40 P/1015.

Lisa

I've often joked I'll need a holiday to get over the festive season but this year I really will. Somehow I've ended up running the Diddly Squat Farm Shop and it has been hectic. Throughout December our own home-grown produce and the things we sell on behalf of neighbouring farms have been flying off the shelves. Admittedly there are not many shelves and they're quite small, but as soon as we fill them they're empty again. We're a tiny outfit – a bit like a car boot sale without the car. People come expecting Diddly Land, only to find basically a shed. Most are good-humoured even when their cars have to be tractored out of the muddy field that passes as a car park.

Behind the scenes baubles need to be ribboned, fudge needs to be ribboned. Actually the Diddly shop crew and I need to ribbon almost everything in between unloading vans arriving from local producers and welcoming flurries of new customers. We now know how Santa's elves must feel.

I'm guessing people who enjoyed the *Clarkson's Farm* TV series like the idea of giving farm-themed presents. Still, I struggle to explain why some of our products are popular. For example Jeremy 'invented' (it's always his idea) a range of body-odour candles inspired by Gwyneth Paltrow's This Smells Like My Vagina candle. I say range – there are two. The one I call Smells Like My Bullocks (the actual title is ruder and nothing to do with farm animals) was our top-selling gift until his Smells Like My Christmas Balls candle hit the shelves. By the way, don't be deceived by the names: the candles give off a lovely fragrance. We also sell Cow Juice (milk), Bee Juice (honey), cheeses, chutneys, jams, tea towels, travel mugs and chopping boards.

I adore everything about Christmas. That said, on the farm it can end up being like any other day because there are livestock to tend to. We were thrilled when our 29 cows arrived because they seemed so gentle and good-natured, not in any way like our belligerent Houdini sheep. The problem is that Jeremy and Kaleb were supposed to build a proper pen, but all they did was put up a bit of feeble fencing while mainly gassing about this tractor and that tractor, then various other tractors, all day long. Our cows have developed a taste for pushing over the fence and escaping. It has become a bad habit and now they are addicted to the thrill of visiting our neighbour's farm.

We were out the other morning before six, herding the beasts for three hours. It's like an early morning workout. As well as running a half marathon you're shaking heavy bags of food. At least it's good for muscle strengthening and toning. I can trot faster than Jeremy so I was lead wrangler. We finally corralled them into their field and Jeremy asked me to go back to get the car. This is when I realised my coat pocket was ripped, probably from the gorse bush I'd crawled through, and the car keys were gone. Retracing my steps, I found the exact hole in the right hedge, in the correct field, and there were the keys. A real-life Christmas miracle right there. Afterwards I took a moment to pause and reflect just how lovely this part of the Cotswolds is. The sun was rising and it lit long streaks of mist over the winter countryside like the brushstrokes of an old master.

Things weren't so pretty when we got home. We have

two red fox labrador puppies, Arya and Sansa – Jeremy loves *Game of Thrones*. In our hurry to chase after the cows one of us hadn't closed the boot room door and the puppies had escaped into the house. Naturally Jeremy said it was my fault that they had relieved themselves all over the kitchen, hall and stairs. I chose to look on the bright side. By another Christmas miracle they had emptied themselves by the time they reached the bedrooms.

It would be lovely if Christmas Eve was like Clement Clarke Moore's poem that begins ' "Twas the night before Christmas, when all through the house / Not a creature was stirring, not even a mouse'. It won't be. If we're not woken by cows on the loose or by the puppies it will be by one of Jeremy's alarms. He recently fitted an alarm to the hen house to stop nocturnal raids by foxes. We've lost as many as 36 egg layers in one night. It has become something of an obsession, so the fox detector is linked to his phone. He had the idea that when the phone beeped he would stealthily reach for his shotgun and blast the vulpine intruder, like Bradley Cooper in *American Sniper*. Of course it doesn't work. The alarm goes off when it's particularly cold, windy or wet or when it just feels like glitching. If I check the phone screen it always shows a dusky image of all the hens silent and fast asleep. Jeremy has to see for himself, so I pretend to sleep while listening to him not silently search for his glasses, knock his water over his bedside books and then begin to swear. If there were a fox it would have long escaped across Oxfordshire, laughing.

Is there a trade union for elves? If not, I may start one for all us Christmas toilers up and down the country. I may even have a T-shirt printed for the farm shop saying National Elf. I'd love it to outsell Jeremy's Christmas Balls.

Lisa's Christmas list

- **Will give Jeremy** – locator tags to attach to his car and house keys. He has no idea where he puts them and days of our lives are wasted looking.
- **Hoping to receive** – a cedar tree.

Kung fu cows

Ants and badgers and kites all seem to manage perfectly well outside in the winter but, for reasons I don't understand, my cows need to be indoors. Which meant I had to build them a barn.

I was quite looking forward to it, being up a ladder with Harrison Ford and some people from the village, and pausing every so often to enjoy some lemonade that had been freshly squeezed by Kelly McGillis. It would be wholesome and fun.

Instead, however, I ended up handing the job over to a company called Fowler & Gilbert. It's run by a man called Lee, whose upper arms are larger than my thighs and who has no need for sleep. He worked from before I was up until long after I'd gone to bed, which is why the entire project was completed in just five weeks.

At £175,000 it wasn't cheap, primarily because it's made entirely from wood, steel and concrete, which, thanks to the demands from engineers building that idiotic Dickensian transportation device from London to somewhere up north, are harder to find and more expensive than gold, frankincense and myrrh.

Still, in early December Lee announced it was done and this meant my cows could be brought from the fields into their new palace. Job one was lining it with straw. There's a machine that can do this but I don't have one, so it meant the bales had to be torn apart using something called a 'fork'. It's even more primitive than a railway line, so the job took two hours and made my back hurt.

And then, having played Tetris with all the gates to create two bays – one for the beefs and one for the veals; they can't be together in case they get up to some nocturnal jiggy jiggy – I had to go out into the fields and make like Ben Murphy and Pete Duel in *Alias Smith and Jones*.

Straight away there was a problem: the cows stubbornly refused to cross the line where the electric fence had been. It wasn't there any more, I'd taken it down and surely they could see that. But they would not be budged and while I was figuring out what to do about this, Storm Barra arrived.

The Daily Mail had been issuing warnings for days, saying, with a liberal dollop of capital letters and much hysteria, that millions would be entombed in an ice hell and that nothing would be left standing. For once they were on the money.

The rain arrived, fat and hard and sideways. It was rain that hurt. Sleety rain. Hail rain. And in seconds I was drenched. There was a river of what felt like glacial run-off cascading down my butt crack and my wellies were full. I wouldn't have been more wet if I'd fallen into the sea.

It was cold too. The thermometer said 1C, but thanks to the 80mph gusts of wind it felt worse. The wind was so strong, in fact, that next month my wheelie bin is having to go on a speed awareness course. It was all too awful for words and every fibre of my being wanted to go inside, make some Bovril and sit in a hot bath for an hour.

But the electric fence was down, which meant the cows, should they take a brave pill, could wander off and I'd have to spend the rest of the day looking for them. So I had to keep going. Despite the awful conditions, I had to march them to the gate at the other side of the field, get them through it and then down the track and into the barn from there.

I knew it wasn't going to work and sure enough, it didn't. I had a plastic bag full of cow treats, which I rattled to let the beasts know that food was in the offing, but as they're pasture-fed, they had no interest in my nutty snacks. One of them, though, did have an interest in killing me.

It's said a cow cannot kick backwards but I can testify to the fact that this isn't true. A cow can kick in any direction. They are like Bruce Lee, only more deadly and determined because, having failed to connect with her hoofs, she went for another tack – putting her head between my legs and then raising it smartly.

This worked well and immediately I was on my knees in the mud with crossed eyes, trying to push the cow away. This, it turns out, is impossible. It's like trying to push Westminster Abbey.

I think this was the lowest point in my farming career to date. Being attacked by a cow while on my knees, in the mud, in a storm, with smashed testicles. And all so I can make beef for a restaurant that I probably won't be allowed to open. I was very miserable.

Happily, though, the attack cow eventually grew tired of trying to kill me and decided, for no reason that I could see, to walk calmly through the gate and on to the track. And the others quickly followed. I didn't look much like Pete Duel or Ben Murphy at this point but I was at least doing successful cowboying.

And five minutes later the entire flock walked into their luxurious new house and began to dine on the delicious silage that I'd been marinating in black plastic for the past four months.

Except one. She took one look at the idyllic surroundings and decided she'd much rather live in the field, up to her armpits in mud and with iced-over eyes.

She took off up the bund, dragging herself up the steep slope by her knees until she reached the summit, where she reared up like an enraged stallion and emitted a sound that was part cow and part werewolf. It was an unholy bellow and once it was over, she ran.

And then it went dark. For an hour I traipsed around with ice in my hair, and mud everywhere else, trying to find it with no light, save the torch on my iPhone. And I'm sorry to say, I couldn't. So I went home, had a bath and then went to the pub.

Kaleb told me I'd done the right thing because chasing a pregnant werecow could cause it to abort. And what's more, he said it wouldn't get far before it thought, 'Wait a minute, I'm all on my own out here and I'm cold, so I think I'll head back to that lovely new barn and join my mates.'

It didn't. What it did do was head back to the field where it had been living and it's still there now. I've tried violence and threats and chasing it with a quad bike. I've even tried enticing it with special pellets made from oil-seed rape casings but it seems not to like these very much.

I've worked with James May for 20 years so I thought I was familiar with the concept of 'stubborn', but this cow has it on an all new level. So that's it. I've given up and all I can do now is wait for the day when a passing badger gives her TB.

In which I'm not opening a restaurant . . .

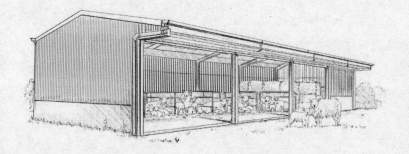

I decided recently to turn an unused lambing barn into a small, wood 'n' sawdust café that would sell good food that had been grown and reared locally. This is the kind of ginghamy, low-impact thing that makes foodies and eco enthusiasts priapic with delight.

And having spent thousands on advisers and landscape architects, I was thrilled that my plan passed muster with the parish council, Thames Valley Police, Oxfordshire county council's transport division, West Oxfordshire district council's drainage division, their environmental health people and, especially, their very helpful business development department.

Sure, the CPRE (formerly Campaign to Protect Rural England) objected and there were 52 people in the area whose real-ale-infused red trousers were glowing puce with nimby-induced rage. But unusually 12 locals had written to the council, off their own bat, to pledge their support. That rarely happens, I'm told.

All of this meant the planning permission would be a shoo-in, so I bought the cows that would produce the beef we'd need, and built them a barn. I also built hen houses that would produce the eggs and my son gave up his job in London to come and make sauces from the chillies I was growing.

Other farmers in the area pledged their support too, which isn't surprising as I'd be paying rather more than what they could get selling their pork and vegetables and chickens to the supermarkets.

So, in the spring, after I'd worked on the lambing barn, fitting lavatories and cladding it in timber from my own woods, and I'd landscaped the grounds, there'd be a ready supply of local food. Brilliant. Except it's not because, as you may have read in the news, planning permission was refused.

The fateful meeting where this happened was held in a pus-coloured room at the headquarters of the West Oxfordshire district council, in Witney. Inside there was a lot of press, an angry rich couple from my local village and, facing the local civil servants and minute-takers, ten councillors from the planning subcommittee.

I scanned them carefully, trying to spot the Liberal Democrats who might oppose my plan because I'd once hosted a car show. But they all seemed fairly decent. One was an old mate. Another was wearing tweed. So I was pretty relaxed when the chairman opened proceedings by inviting the rich angry couple's barrister to explain in less than three minutes why permission for my café should be refused.

I was looking forward to this because he had a badly spelt report containing an idiotic mathematical miscalculation.

It reckoned, for example, the car park I wanted would cover 500 acres, not 0.5 of an acre. But instead of using these hilariously inaccurate facts and figures, representing his onlooking clients, he launched into a character assassination of me, saying my behaviour was 'shameful' and that I had a 'give me an inch and I'll take a mile attitude to planning'. He didn't actually say I smelt like a wee and a poo but only, I suspect, because he ran out of time.

A bit perturbed, I didn't use the full allocation of three minutes to put my case. I just stammered my way through an explanation that farms cannot survive unless they're allowed to diversify and then, mercifully, the floor was handed over to the councillors so that the process of local democracy could begin.

It was horrific. They didn't seem to have any facts to hand and one of them wondered why I couldn't open the café on someone else's farm. Mostly, though, they seemed to be extremely bothered by the fact that the barn was in an area of outstanding natural beauty, not understanding perhaps that it's only beautiful because farmers keep it that way. They also seemed concerned about how much lighting would be needed and how this would affect the night skies. Not as much as nearby RAF Brize Norton does, I thought, but I wasn't allowed to speak.

At one point a sensible younger woman said that people in the area cared more about how many jobs would be created than how many stars they could see, but she was immediately shut down by the chap to her right who said tetchily he liked to watch *The Sky at Night*. Right. I see. Because you like watching a TV show about galaxies I shouldn't have permission for a café.

As the time wore on people made important-sounding noises until eventually the planning officer said she wouldn't recommend permission be granted. And then a man called Phil, who was obviously some kind of local government boss, made a speech that built to a crescendo of fury, saying that the café would cause 'great harm' to the area.

As that tidal wave of misguided moral righteousness swamped the room, he called for a vote, won it easily and just – but only just – stopped himself punching the air with delight.

Of course I get where he's coming from. Nobody has ever left university saying, 'I want to run a subcommittee at my local council.' So I understand that he's constantly disappointed with how life has turned out for him. As a result he will see this vote as a tremendous victory for the downtrodden little man who took on, and beat, the man off the telly.

I don't doubt the rich angry couple from the village are happy too. Doubtless they're in their onyx pool now, celebrating the win with some prosecco and a few cheesy Wotsits.

But what this unholy alliance actually did was say to the thousands of struggling people who farm in a national park, or an area of outstanding natural beauty, or the Norfolk Broads, that they may not turn unused buildings into cafés or gyms or office space.

It's nearly impossible to make money from farming these days and in recent years farmers have relied on government grants to keep going. But those grants are being phased out and we have been told by the government that to survive we must diversify. And now local government is saying we can't. That has to be addressed, and soon, so that councils are encouraged or even ordered to cut farmers some slack.

What's more, everyone has spent the last six months moaning about how I must build a car park to stop visitors to the farm shop from parking on the road. And now the council has said I can't.

The day after the meeting I was swamped with offers

of help from other councillors on other planning authorities who say the council's report was full of holes and that the decision was absurd. Lawyers have been in touch, out of the blue, to say I have a strong legal case. I also received kind emails from Roger Daltrey and the chair of the health, wellbeing and social inclusion working group of the Cotswolds National Landscape, who says she's 'deeply discomfited' by the decision. Naturally the local farmers, hoteliers and publicans are devastated, because the people the café would have attracted won't now be coming.

Of course, eventually, this unholy mess will all be untangled and the decision will be overturned, but in the meantime I'm going to have to keep everything going by living in the murky, grey area of legal loopholes and cunning wheezes. I'm good at this, though. It's all I learnt at school, really, getting round the rules rather than abiding by them.

More immediately, though, I have a cow in the barn that is due to give birth to her first calf. A calf that, thanks to Phil's planning subcommittee, will now be sold at a loss to a supermarket, where it will sit on a shelf until it's an hour out of date. And then it'll be thrown away.

Bovine blues

You might think that farming is a kaleidoscope of ginger beer, butterflies and leaning on gates as the sun sinks gently into the dandelion-seeded western hinterland.

Well, last week it wasn't like that at all. It was boring. I'd had to talk to an insurance man about a fat woman who'd come to the farm shop wearing Crocs and then fallen over. And then, when my eyes were tired from rolling, we moved on to how much cover I'd need in case some baddies hacked my computer system and held me to ransom. Much like they did to KP Snacks at the end of January.

After this there were endless discussions with men in suits about how I can put a café in my unused lambing barn when the council has said I can't.

And then Cheerful Charlie, the land agent, turned up to discuss the future of farming grants, the stratospheric cost of fertiliser and why an entire field of oil-seed rape had simply not grown. And what we'd plant there to replace it. I understood some of the words he used but not enough to know what he was on about. Apparently I'm going to try and grow fish oil.

This is the reality of wintertime farming. You want to get out there and mend the broken dam with a big and exciting machine, but you never can because there's always a meeting. And there are always forms to fill in.

Take my cows as a prime example. I thought they'd be fun to have around but the government's cow police does everything in its power to make sure they aren't. I have to fill in a form if I want to take them for a walk, or sell them, or eat them. And every six months I'm forced by law to test them for diseases.

You don't have to do this if you have a cat, or a rabbit. I have three children, for crying out loud, and once I'd registered their births I could do pretty much what I liked with them. Apart from serving them up for Sunday lunch. What's more they can come and go as they please. I think one of them is currently in Los Angeles but I couldn't say for sure. Only if she were a cow would I be certain.

Four of my cows gave birth to heifers (girl calves) last week, which meant I had to spend an hour informing the cow police, and then I had to go outside, in the snow, to give them huge plastic earrings. These are identifying tags and they'll be in the poor animals' peripheral vision until the day they die.

Fitting these tags is not easy because you must first of all climb into the individual pens where the calves live.

This means scaling a 5ft-high metal fence that is absolutely covered in their mothers' faeces. And then, when you've fallen into the pen, because faeces is very skiddy, the mother cow will decide that you are a threat to her baby and that it'd be best if you weren't alive any more.

So now you're facing down the irate mother cow while working on a plan to get the calf into a corner. If you can achieve this, and it's about as easy as organising air, you must then leap over the calf so that you end up straddling her neck. Then with one hand you hold her head back while with the other you reach for the ear-piercing tool.

You then use your other hand to take two correctly numbered tags – one for each ear – from a bag containing 2,000 tags. This means retrieving your reading glasses, which fell out of your pocket when you fell off the fence and are now on the other side of the pen, being shat on by the increasingly distressed mother.

Have you ever tried walking when you have a distressed calf between your legs? You may think it's easy because the calf is so young and wobbly. I did too. But a calf is still several hundred pounds of pure muscle so it's not easy at all, especially when you absolutely must not spill any of the 2,000 tags from the bag, or drop the iodine bottle that you're holding in your fourth hand.

Eventually I got the cow into position – although

when I say I, there is some televisual evidence to suggest it was Kaleb – and then I had to position the ear-piercing tool on the ear, taking great care to ensure that when I squeezed the handles I didn't accidentally sever one of the two gigantic blood vessels in there. This is really hard to do because the calf's head is very much a moving target. It's like shooting at an epileptic elephant shrew.

Terrified I'd attach the tag to the poor thing's eyelid or forehead or, worse, Kaleb's penis, I was not speedy. But after just a few hours all four calves had their iodine-soaked, cow-police-issue identifying tags, which meant they could come out of their own pens and play with each other in the main barn. It was a wonderful moment. The sort of thing that I live for up here among the forms and the faeces.

But it was short-lived because immediately the vet arrived, under instruction from the cow police, to check my entire fleet for TB. This is a disease passed on to cattle by badgers, and if they have it they have to go 'down the road'. That's sad for me because I love my cows and it's sad for you too because you have to foot the bill.

I'm told that because I live in a high-risk area there's a 'good chance' one or more of my cows will have to be executed, and what makes that even harder to bear is that if one of my pregnant cows is infected, they have to kill her unborn calf as well.

It's a distressing job at the best of times, trying to get the cows into the so-called 'crush', which holds them in place when they need to be examined for some reason. But it's doubly distressing when you're trying to get them in there so they can be tested for a disease that'll result in their death.

One was so heavy with calf she wouldn't fit, so we had to try and hold her in place while the vet did his thing. Which was a bit like asking Susan Boyle to do a c-section on a *T. rex*. And then we had to turn our attention to Genghis, the attack cow, who so far has kicked Kaleb twice and smashed my testicles once. I don't know how that went, as I suddenly remembered I was due at a lunch. I also don't know yet how the test went. All I can do in the meantime is hope and pray.

Of course you might be wondering why I don't minimise the risk of my cows catching TB by reducing the number of badgers on the farm. Well, there's a simple answer to that. The badger police won't let me. And if I do it anyway, I'll have to fill in a form saying that the desk sergeant has my watch and belt and that I'll want them back when I'm released in about 2037.

SPRING

The big fertiliser dilemma

I have watched two box sets on Netflix recently. One was about a conman and the other was about a con-woman, and in both cases I was left thinking, 'How does anyone have the brain capacity to live a life like that?'

In the first, called *The Tinder Swindler*, there was a disgusting weasel from Israel who kept a string of European girls on the go, and then somehow convinced them to give him all the money they had, and a lot of money they didn't have, so that he could buy awful clothes and use private jets to get from shag to shag. And all the time I was thinking, 'How come he doesn't get muddled up from time to time?'

Then there was *Inventing Anna*, which was about a girl who convinced a bunch of extremely stupid art and fashion people in New York that she was a German heiress, and as a result they lent her their boats and jets and hotel suites. And again I was thinking, 'How much spare capacity must this girl have in her head to keep all those plates spinning?'

Imagine what life would be like if you had to

remember, when you ran into someone at a party, how much you owed them, what lies you'd told and what excuse you'd given for not paying them back. It's impossible even to contemplate. Me? I sat opposite a very funny chap at lunch the other day and secretly had to google clues from our conversation even to come up with his name.

I therefore had great respect for the conman and the conwoman, in the same way that I have great respect for the businessman who sits on lots of different boards, and bosses who have to run their companies while dealing with the mental health issues of every single employee. It must be exhausting.

I get worn out if I'm wiring a plug and someone asks me for the time. I'm never able to reset my head so there's a blank piece of paper inside it. It's always full of doodles from the last thing I was doing, and phone numbers and notes for columns I must write later in the week. I wonder if politicians have the same issues when a new crisis comes along. I suspect they do.

Anyway, this inability to focus on complex problems is one of the (many) reasons why I'm turning out to be a not-so-good farmer. I simply don't have the mental capacity to make rational decisions when more than one factor is at hand. Yes, when I'm hungry I can decide easily to have a pork pie, but if there's a piece of tongue in the fridge as well? Well, that's me paralysed by indecision for an hour.

And in farming right now, it has gone way beyond pork pies and tongue. Because of rising gas prices caused by all sorts of world events, fertiliser prices have shot up from about £250 a tonne to three or four times that. Many farmers are therefore thinking about using less on their crops, which will reduce the yield.

That sounds bad, but they reckon that because 30 per cent of the world's wheat and barley comes from Russia and Ukraine, the price of what they sell for will rise and therefore compensate for the miserable quantities that will result from using less fertiliser.

This 'grow less but get more for it' philosophy may be right, but what if every farmer on earth went for the lower-yield option? There wouldn't be enough food to go round. And pretty soon people would be beating their elderly next door neighbour over the head with a baseball bat to get at the contents of her bread bin. And murdering the milkman so they can lick the float in the hope that a bit of last week's load spilt somewhere.

And I have an even bigger issue to wrestle with. I* bought my fertiliser early last year when prices were fairly low. If I sold it now, I'd make a profit of maybe

* *Obviously by 'I' what I mean is 'Cheerful Charlie, the land agent'*

£30,000. But then I'd have none to put on my crops. How much would this affect yield up here in the brashy stratosphere of north Oxfordshire? And what if the war ends tomorrow and everything returns to normal?

If I sell my fertiliser, then I'd be betting on the war ruining the harvest in Ukraine. Which means I'd be sitting here, praying the conflict and all its attendant awfulness carries on right through the summer. I'd be a war profiteer and I couldn't do that, so what can I do?

I find myself spending hours in my grain barn trying to figure this stuff out but everything just swims round and round like I'm stuck in a Federico Fellini movie where there's a priest and a crow and some clouds and a circus, and I haven't a clue what's going on. And I can't concentrate because someone is playing a violin backwards.

I'm rooted to the spot, incapable of making a decision. Fight or flight? Flee or wee? Pork pie or tongue? I thought farming would be mostly chewing on bits of grass while leaning on a fence, not this. Not playing geopolitics. You need that guy from *The Tinder Swindler* to do that. Or a team of analysts called Brad and Todd, who offer visitors 'a water'. And I don't have either. No farmer does, and that's something you should worry about.

Brexit caused staff shortages at Britain's already woefully small number of abattoirs, which means that instead of putting their hogs into the system, pig farmers are being forced to kill them and throw the carcasses away. Soon this will cause bacon to become more expensive than swan.

Then you have dairy farmers, whose money is delivered by an army of snails. The cash flow situation is now so terrible that they are having to sell their lady cows for meat. Which means that soon milk will become more dear than champagne.

Then there's the global youth movement that has decided that badgers are lovely, which means thousands of meat cows are dying needlessly from tuberculosis, which means your burger will soon cost more than your house.

And now comes this terrible and stupid war, which is going to cause bread, pasta and vegetable oil to become more expensive than gold, frankincense and myrrh. And it's no good saying, 'Oh, we will just get our food from abroad', because farmers over there are in the same boat.

And because we tried that once before, in the 1930s. Just before the U-boats came along and damn near starved us.

The fact is, you can live without sex. You can live without box sets and clothes and cars and holidays and even houses. But you cannot live without food. Oh sure, you might manage a day or even two, but after three the hunger will take over and you'll do whatever is necessary to feed yourself. And more than what's necessary to feed your children.

I wonder if our leaders realise this. Or are their heads still too full of all that Partygate nonsense to concentrate properly?

What not to eat

We all accept that life is the most precious thing and as a result we would burn our beliefs to save a friend, and ourselves to save our children. Some go further still, making themselves pasty-faced and weedy by only eating plants, because they don't want to be responsible for the death of an animal. And that's where things start to get complicated. Especially if you want to be a farmer.

Last weekend my friend Reggie, who owns the Cotswold Wildlife Park, invited me over to see a rhinoceros that had just been born. And as I stood there looking at its fantastic mother, I found myself wondering how on earth anyone could kill such a magnificent thing.

I appreciate, of course, that the poachers are poor and that the only way they can feed their children is by feeding the idiotic Chinese belief that horn makes you horny. But I like to think – and it's easy sitting here at my quartz-topped kitchen island – that if I were in their sandals I could not and would not pull the trigger. No matter how poor I was.

So how then can I be a farmer, because what I do all

day is look after my cows and feed them and keep them warm so that one day in the not too distant future I can murder them for money? No. Hang on. It's worse than that because, actually, I pay a hitman to murder them for me.

It's weird. I truly love animals and especially birds. But I also love roast beef and chicken and foie gras and baby sturgeon and oysters and pork and ham and bacon. And get this for added confusion: while I will happily shoot a partridge so that I can eat it with some sprouts and a bit of mashed potato, I won't shoot an equally delicious woodcock. Why? No idea.

I've long held a belief that we only really care about animals that are one of three things: cute, magnificent or delicious – Attenborough, for instance, rarely covers the stickleback – but actually, it's more muddled than that.

I would happily shoot a grey squirrel – and have – but could not even break a whisker on its red cousin. I wouldn't hurt an otter but would happily stamp repeatedly on the head of a badger. And even though the farm's deer are laying waste to my trees, I really struggle with inviting someone over to walk them 'down the road'.

I think all of us are similarly conflicted. Even the most hardcore peace 'n' love vegan will smash a wasp over the head with a copy of the *Socialist Worker* if it's being annoying. They will also spray their holiday hotel room with mozzie killer before they go out for dinner. And even if they are so fanatical they don't do either of these things, I bet they'd be happy for someone to shoot a crocodile if it was eating them at the time.

All of which brings me on to the mouse that Kaleb found in the frame of my seed drill last week. As a proper farmer he wasn't the slightest bit bothered and, as the weather was closing in, told me to get out there immediately and start planting the spring barley. But I couldn't go out into the fields and unfurl this enormous machine – it's like an oil rig, only bigger and more complicated – knowing that a sweet little mouse was in there. And that in all probability it wouldn't survive.

Kaleb was staggered by this and pointed out that the whole farmyard is littered with mouse and rat traps. I was stumped because he had a point. But I still couldn't do it and so, to a chorus of tutting noises and 'oh for God's sake' exclamations, I found a length of hosepipe, which I gingerly inserted into the frame until, after a few moments, the little creature fell out and dashed for the nearest bit of cover. Which was under the back wheel of my tractor.

I got down on my hands and knees and I could see the poor little thing, cowering in the tread pattern of the tyre. I then spent some time assessing the situation before coming to the conclusion that Kaleb had reached several minutes earlier – there was no way in hell I could get it out of there.

'Now what are you going to do?' Kaleb asked impatiently. 'I'll tell you what I'm going to do,' I replied. 'The best hill start you've ever seen.' And so I climbed into the cab, engaged first on all of the 42 gearboxes and then very gingerly pushed the button that would raise the 6m seed drill off the deck.

Straight away there was a problem. Because as this 3.5-tonne machine rose from the ground, it pulled the tractor backwards about nine inches. A sickening nine inches. A crunchy nine inches. There was no way the mouse could have survived and I was white-faced with horror and guilt. And what made me feel more awful was that Kaleb was standing there, shaking his head and saying, 'You are so not a farmer.'

He had a point, though. The mouse would have been eaten that night by an owl and now it was carrion for the red kites, so all I'd achieved by faffing around was to delay getting my spring barley in the ground.

And so, with a heavy heart, I let the clutch in and set off knowing that after the tyre had done a quarter of a rotation, a small red splodge would become visible. It never got that far, though, because after I'd moved just a foot, the mouse shot out and set off at what appeared to be 2,000mph across the farmyard. And into the path of Kaleb's brother, who'd just come round the corner carrying some hay for the cows. Seeing a mouse, he did what any farmer would do. He lifted his size-12 boot, stamped down hard and . . .

Missed.

I was delirious with joy but the truth is that I've not felt more like a townie since I started this farming malarkey. Worrying about how many mice and rats you kill in Mudfordshire is like a lorry driver worrying about how many flies have splattered into his windscreen. Or a rambler losing sleep over how many earthworms they squidged that day while establishing their right to trespass.

All of us are different when it comes to animals. Some are Kurt Zouma and some are Chris Packham. And then you have people like me who'd happily nurse a baby hedgehog back to robust good health while cooking a hearty stew.

Since I started farming I've been permanently conflicted, especially yesterday, when I spent a whole day rearranging 120 yards of hedge to make it more friendly to nesting birds, before heading off in the evening to waste the pigeons that have been eating the bits of rape that haven't already been devoured by the flea beetle.

Vegans would say I was wrong to do that because

even a pigeon's life is precious. But if I hadn't, I'd have no rape oil to sell, which would cause many people to use palm oil instead. And that would be very bad news for the world's orangutans, whose lives, in my book, are more precious than a flock of airborne rats.

I guess that's what I need to remember. All the animals are equal but some are more equal than others.

The answer's still 'no'

Many years ago, the Conservative Party realised that a person who owned their home was more likely to vote Tory. It's why Mrs Thatcher gave people in council houses the right to buy them. It's also why, last year, the Tories announced that 300,000 new homes would be built, and that red-trousered village nimbys would be stripped of their right to object.

But then, in a forehead-smacking moment of horror, they realised that almost all the nation's red-trousered village nimbys were Tories, and that if they were no longer allowed to object to every damn thing, they'd quickly switch allegiance to another party.

It's one of the main reasons the Tories lost the Chesham & Amersham by-election last year. Voters in this historically deep-blue area were terrified that developers would be given carte blanche to carpet-bomb the area with awful executive homes full of ding-dong doorbells and baggy-knicker curtains. And they voted for the Liberal Democrats instead.

So then the Tory leadership had a problem. If it allowed

developers to build houses that would turn Labourite benefit fraudsters into rosette-wearing, *Last Night of the Proms*, homeowning Conservatives, it'd turn all the red-trouser people into liberalists. It'd be robbing Keir to pay, er, whoever is the leader of the Lib Dems these days.

Which is presumably why last week, in the Queen's Speech, we were told the Tories had a new plan. They would allow local communities and villages and even individual streets to hold referendums on whether or not developments should take place. Because nothing bad ever comes from a referendum, does it?

They hope this will lead to what they call 'gentle densification', by which they mean that a street full of semi-detached houses could become a street full of terrace houses, with children playing hopscotch on the pavement while Dad mends his Cortina and Mum bakes cakes for the new neighbours who are now living in what used to be her back garden.

I must confess that I don't understand all the details, but I do see that the idea, on paper at least, has legs. The country gets more houses, which means fewer Labourites. The developers get to build these houses, which means they make money. And the red-trouser people get to vote on what goes up and what doesn't. Everyone wins.

Except in practice I can't see it working at all, because the red-trouser people will always vote for things to remain exactly as they are. They certainly won't be swayed by the promise of more low-cost housing, because that's the last thing they want. Poor people? In their village? Noooo.

This, of course, is why we have local planning authorities. They're there to balance the views of the developers against the views of the locals and then decide which argument holds more water. That's obviously the best solution to the problem. Except that doesn't work either.

Everywhere you go, you see hideous new office blocks and you wonder: 'Who the bloody hell thought that was a good idea?' You see awful housing estates on flood plains, and barns that have been converted into mock-Tudor mansions, and in London things are even worse. Someone is building a hotel on Hyde Park Corner, and every time I drive past it, I'm filled with an urge to murder the people who gave such a monstrosity the green light.

In my mind, freemasonry plays a part in many planning decisions. You put on an apron and roll up your trouser leg, and Bob's your uncle: you can have a new conservatory and let's tear up that speeding ticket while we're at it.

Of course, I'm not suggesting for a moment that my local planners are swayed by shenanigans at the lodge. But this means I have no clue at all how they arrive at their decisions. I think they lick their fingers, hold them up to the wind and then say: 'If the applicant is Jeremy Clarkson, the answer's no.' I have no idea what I've done to upset them, but it must have been something pretty terrible.

I recently applied for permission to turn an unused lambing barn into a small restaurant. Naturally, the red-trouser people in the nearest village objected, because that's what red-trouser people do. But I put forward sound business arguments, which were backed by the council's own business development people. And I lost.

I then asked for permission to build a small farm track that would reduce the need for me to drive my tractor on the road, cut my fuel bills and improve the environment. And they said no to that as well.

In recent months everyone has moaned that visitors to my farm shop are parking on the road and making it difficult for locals to get past. So I asked for permission to build a car park. And even though it would be right next to a caravan site and just a few hundred yards from a very large travellers' camp and a field where the army practises helicopter landings, they said no, because 'it's an area of outstanding natural beauty'. They didn't even put that one in front of the planning committee.

In the national planning policy framework, planners are told in paragraph 38 to 'approach decisions on a proposed development in a positive and creative way'. But that sure as hell didn't happen in my case. And, so far as I can tell, it's not happening anywhere. The planning system we have now is completely broken. It's slow. It's expensive. And it's too weird.

The new system proposed last week may or may not be better, but either way I already have a plan to exploit it. I'm going to have a referendum in my village on whether we should demolish the house of the red-trouser enthusiast-in-chief. As he's made a number of enemies with his objections in the past, I suspect he'll lose. Which would be hilarious.

Feed the world

I do tend to go through life with a general sense that everything will be all right in the end. Yes, we are told every 20 minutes that soon the Earth will be a superheated ruin that's no longer capable of supporting even bacterial life, but I continue to run seven cars, six of which have V8 engines, because I reckon that in the nick of time a Munich-based boffin will invent a giant space-based vacuum cleaner that will hoover all the unnecessary carbon dioxide out of the atmosphere and make everything normal again.

I had the same attitude with Covid. Of course it wasn't going to wipe us out, because somewhere in Germany there'd be a scientist in a room full of pipettes and Bunsen burners who'd invent a vaccine. And so it turned out to be.

Financial crash of '08? Yup, I did a bit of running round in circles back then, thinking that all my savings would be consumed by the invisible and unfathomable fugazi that is Wall Street, but soon there was quantitative easing, and a deal with China, and by the spring of '09 my snout was back in the fish roe.

This Ukraine business, however, is causing me to have a few chin-scratching moments of despair. I don't pretend to be an expert in geopolitics any more than I pretend to be a farmer, but I really think the world has slipped into a pair of margarine trousers and is now hurtling down a well-watered slide into the pit of hunger, misery and death.

Let me run you through my thinking. The conflict and the sanctions and all of the other flotsam and jetsam that hurtle round a war zone have caused gas prices to skyrocket. You know this, of course, because it now costs a million pounds to heat your house and £20,000 to cook a lamb chop. I know it, too, because chemical fertiliser has gone from about £250 a tonne last year to about £1,000 a tonne now.

Naturally, because you don't need fertiliser, you don't care. But you should care because soon you're going to go to the supermarket and all you'll be able to buy is an out-of-date copy of *Auto Express* magazine and maybe 20 Benson & Hedges. And then, on the way home, you'll be murdered.

The problem is that next year many farmers will decide that, because of the costs involved, they'll use less fertiliser. Some will doubtless try to use none at all. Others will try to use cardboard or lawn clippings or faeces instead. Either way they will produce less food. Some farmers – I know of three in my area alone – have already decided to fallow their fields next year and grow nothing at all.

And this is not just happening in the UK. It's a global phenomenon and it could well result in there being maybe 20 per cent less food in the shops than is necessary. That's bad. And then it gets worse because, between them, Russia and Ukraine grow more than a quarter of global wheat exports. They are also responsible for about half the sunflower seeds we use, which is why, already, sunflower oil is being rationed in Britain.

So, thanks to the war, we lose a lot of the grain we need, and then, due to the cost of fertiliser, we lose 20 per cent of what's left. Prices are already going up, not by 7 per cent or 10 per cent but by a massive 37 per cent. And the World Bank says it won't stop there. They call it a 'human catastrophe'.

Politicians say they are 'monitoring the situation', which means they aren't doing anything at all, but one day they will have to because while people can live without heat or clothing or even sex, they cannot live without food. Hunger makes people eat their neighbours.

Or move. And surely that's what must happen next. It's said that nearly a third of the wheat Ukraine grows goes to Africa, and it won't be getting any this summer. Nor will Africa be able to afford mine. Not at £300 a tonne. Anyway, what I grow is going to be needed here.

So what do you do if you are in Africa, or the Middle East for that matter, and there is literally no grain? Sit around waiting for Midge Ure to fire up his Nokia and call Bob Geldof? No, you're going to up sticks and move to the only haven that's remotely accessible: Europe. We've seen a lot of migration in recent years but I suspect that soon we'll realise that what we've had so far was only a trickle.

So now the streets of Europe are filled with hungry and desperate immigrants claiming 40 quid a week from the government and finding that it isn't even enough for a loaf of bread. Add them to the poor indigenous folk fed up with choosing between heating and eating and that's when things risk turning really ugly.

It's hard to see what on earth can be done to stop it happening. The British government could take a lead and force farmers to farm their land, with grants to pay for the fertiliser and nationwide clapping every night at eight. But that isn't going to happen for a couple of reasons. First, the British government is run by Carrie Johnson, who thinks the countryside should be for badgers and not for growing food.

And second, the rest of the government (and the fourth estate, if I'm honest) is currently consumed by whether a slice of cake can turn a work gathering into a party and simply isn't paying attention.

I get this, of course. They're like me, assuming that a German with a Bunsen burner will come to the rescue, but this time I can't see that happening. The war has chopped off a quarter of the world's grain exports and caused gas prices to skyrocket, which means farmers in the West can no longer afford to feed their crops properly. Less food and massively higher prices are the inevitable consequence, and the result of that is hunger and many arterial blood splatters across the fridge-freezers in your local Iceland.

Prince Charles will tell you that the Arab Spring uprising was caused by global warming. Indirectly he may be right, but the direct cause was a sudden jump in food prices. And the world is still feeling the effects ten years later.

This time, though, it'll be worse. And then we'll get the right-wing, anti-immigration parties leaping onto their soap boxes and blaming the EU, which will cause Europe to fragment and then the world's last bastion of liberalism and common sense and decency will be broken.

Which is exactly what Putin wants. Sure, the war in Ukraine may result in him gaining only a tiny bit of land in the east of the country, but beyond that it could well destabilise Europe for years. Unless, of course, none of that happens and the continent is saved, hilariously, by a German. In which case we could all go back to worrying about whales and global warming.

SUMMER

The grass is greener

I was coming home from the pub the other night when I saw a car parked at the side of the road, in the middle of nowhere, next to what I call 'the Big Wood'. And as there have been reports in recent weeks of poachers in the area, I thought it might be prudent to stop to see what was what.

So I pulled up alongside the car and inside was a young man who was using one hand to smoke a large joint and the other to, how can I put this, frantically pleasure himself. Lisa and I had time to exchange a smile before the young man realised we were there, stopped what he was doing and lowered his window. 'Oh my God,' he said. 'You're Jeremy Clarkson.'

Now at this point a normal person would have made all sorts of mumbled excuses about how they'd pulled over because they'd had an, um, itch and had decided, while scratching it furiously to, er, light a perfectly legal roll-up. But instead the young man said coolly, 'I wish we could have met in different circumstances.'

I still chuckle when I think of this exchange. Which is

good because, in other news, my rape's gone wrong. Which will come as no surprise to those who say it was idiotic to try to grow it in the first place.

A few years ago oilseed rape became such a useful rotational crop for farmers that when you flew over the countryside it was like England had become a yellow and pleasant land. And not only was it good for the soil but it could also be used to feed cows and make bio-diesel and extremely healthy vegetable oil.

And then along came the EU, which said that the neonicotinoid seed coating used to protect the crop from insects was seriously buggering up the bee popula-tion, and banned it. Rightly so, in my view.

Without neonics, of course, the crop was at serious risk of being eaten before it had even had much of a chance to get out of the ground, and as a result many farmers started growing something else instead. But I persevered and two years ago lost an entire field to the flea beetle. Hours of work. Thousands of pounds. Wasted.

I was told, however, that the pesky beetle could be defeated if I planted the crop earlier than usual. Or later. The only thing I couldn't do is plant it at the correct time, which, for someone who is pathologically punc-tual, is quite difficult. And it hasn't even worked. Because in the one rape field I can see from my kitchen window, it looks as if Jackson Pollock's been round.

Infuriatingly, right next to the field in question, my neighbouring farmer's rape crop is a strong and smooth yellow blanket. And way off in the distance there are other fields that are doing well too. This is giving me serious, lip-curling farm envy. It's not just that they're doing better than me, which is bad enough. It's that they can see, very clearly, I'm doing so much worse. All they have to do, if they want a laugh every morning, is open the curtains.

I can't think of any other job that's like this. Plumbing is hidden. So are surgery and accountancy. But farming's out there in full view. And what my wonky field says is: 'Hey, I'm useless at this.'

Everyone is going to assume I drilled (seeded) that field and forgot to push an important button in my tractor or loaded the hopper all wrong. But I didn't. Kaleb did.

So I consulted him and he had an answer straight away. He says it has all been eaten by pigeons, but while there have been many more of the aerial rats knocking about this year than is usual, it doesn't take long to work out that his argument is flawed. Because why would they only target my crop and not the crop in the very next field?

'Ah, that's your fault,' said Kaleb. 'Because you insisted on having wildflower runs in the field, which are like airports for the pigeons.' Kaleb doesn't like my work to increase insect life. He thinks it's a waste of money.

Cheerful Charlie, the land agent, doesn't agree, however, that my eco-friendly beetle runs are the issue. He says that my neighbour used a different type of oilseed rape and that plainly the pigeons preferred mine. So now my neighbour can gloat about that as well.

I only learnt last year that there are different types of rape. (That's not going to look good if a tabloid takes it out of context.) This is because I met a man who asked me to try his variety. It was one he'd developed and, he said, he therefore owned it. I was puzzled by this because how can you own a type of plant? It'd be like me saying that I own daffodils.

Anyway, we've bought the wrong sort of rape and soon, no doubt, they'll divert planes coming out of Heathrow so that passengers can have a good laugh at Jeremy Clarkson's continued failure.

And that's before we get to my grass fields, which I need for the cows and which have no grass in them. This baffles me because when I stand in a field that's entirely green I always assume it's grass. But somehow it isn't. It looks like grass until you get down on your hands and knees and break out a magnifying glass, and then you discover that it's mostly just weeds that have the nutritional value of cardboard.

I suggested that maybe we should use a bit of fertiliser to encourage the grass to get cracking but that caused everyone to laugh at me again because a) fertiliser will also encourage the weeds and b) it currently costs more per gram than cocaine.

I genuinely thought when I started farming that to grow crops you put seeds in the ground and then sat back as weather caused them to grow. But it's not like that in reality because weather usually causes them to die. And on top of that you've got price squeezes, scientific breakthroughs, environmental pressure, new legislation, new technology, wars, pandemics and a government, in this country at least, that seems to have no understanding at all of what's required.

Meanwhile, you're expected to know how deep the seeds should be planted, when you should spray them with fertiliser and what other plants to grow underneath them to help the soil.

And while you're dealing with all this science, you have to bleed the brakes on the tractor with one hand while shooting at a huge herd of pigeons with the other.

And if by some miracle you do manage to produce something at harvest time, you're lambasted by the water authorities for polluting their streams, criticised by environmentalists, abused by supermarkets, hated by the nation's rambling enthusiasts and given absolutely no help by the government, which has it in its head that farmland is nothing more than a handy spot for disaffected city youths to come out to at night for a doobie and a Jodrell.

Dambusted

Every autumn we are treated to the distressing spectacle of people, in places like Worcester and Doncaster, hauling soggy three-piece suites and ruined fridge freezers from their flooded houses. And every year farmers are blamed by the likes of the eco-journalist George Monbiot for not managing the soil properly.

That's why, a couple of years ago, I hired a very large digger and spent a week roaring about in it, making dams and boggy places. My thinking was simple. If I could hold the rainfall up here, high in the Cotswold Hills, it wouldn't cascade down into the local village and ruin everyone's red trousers. Plus, I'd have some nice ponds.

And far away in Norfolk an 84-year-old farmer called Paul Rackham plainly thought much the same thing. So he cleared away all the weeds and brambles on the banks of the Little Ouse, a small river that runs through his land. This caused the water flow to slow down and created a lovely spot for swans to come and live.

And because of this his company has been fined £17,000, ordered to pay £49,000 in costs and forced to spend £400,000 doing remedial work. Welcome to the world of farming. You try to do your best and someone with a clipboard and adenoids sends you a bill for nearly half a mill.

It seems that a government spy visited Rackham's farm to take water samples. He noticed the river was deeper than usual and, possibly because he doesn't have deep-water training, he had to abandon his sampling. A month later more spies visited the area to see why the water was deeper and found the banks had been tidied up and raised.

Yet more spies arrived in a fleet of hired Vauxhalls and somehow, in the new banks, they found evidence that water voles had been living there. In my experience this is an inexact science as water vole holes look much like any other sort of hole. But, anyway, they knew a water vole community had been destroyed along with a number of unnamed invertebrates and shrimps. So they spoke to Rackham, who immediately stopped the work.

He explained that he'd had consent from the Environment Agency to make flood defences in the past but admitted he did not have a permit to do so on this occasion. And now, as a result, he's out of pocket to the tune of £466,000.

I've said before that people only really care about creatures that fall into one of three categories: cute, magnificent or delicious. And the water vole is definitely cute. It may even be delicious as well; I don't know. Foxes and buzzards seem to think so. One thing's for sure, though, it is extremely rare.

In my lifetime numbers in the UK have fallen from about eight million to under 200,000 today. And most of those live in Glasgow, for some reason. So it needs some protection from farmers and landowners who behave recklessly. But a fine of £17,000? I can't even imagine how fast you'd have to drive a car to get clobbered that hard.

I was interested in the story for another reason, though. It's because I've been busy these past few months planning to repair a dam on my farm. Installed in about 1368 by Geoffrey Chaucer, and made from wode and oak, it's disintegrating and I fear the lake it was built to create will soon drain at high speed into the living rooms of all the red trouser people in the village.

Things aren't going very well with them at the moment. They complained that people visiting my farm shop were parking on the road, and when I applied for planning permission to solve the problem with a car park, they objected to that. And now the council has turned me down.

In some ways I'd quite like the dam to break so all their trousers get soggy, but for the sake of the trouts that live in the lake and the ducks that live on it, I must effect repairs. However, I just know that when I apply to the Environment Agency for permission, two things will happen. The red trouser people I'm trying to protect will object, and the clipboard people will find evidence of water voles. Or bats. Or newts.

And then the real problems will start. You see, I planned to finance the lake restoration programme by catching crayfish and then selling them in the shop, either as a glorified prawn cocktail or, in the winter, as a chowder. Clever, eh?

You go down to your own lake on a lovely summer's evening, haul in a net full of delicious morsels and then sell them to passing families as a healthy snack. Except I can't do that because this isn't a free country.

The problem is that the crayfish I have are American, which have done to the British variety what the grey squirrel did to the red. They are therefore labelled as an invasive species, which means the government is forced to spend millions of our pounds employing a team of people to make and apply rules about what can and cannot be done with them.

So I have to have a licence to catch them in the same way that pilots have to have a licence before they can fly a helicopter. They therefore want to know my name and address and how big the lake is exactly. And what sort of water it contains. And whether it's still water or flowing water. And the precise location. And whether the site is somewhere of special scientific interest.

Then they want to know what sort of trap I'll be using, pointing out that it must be no longer than 600mm and no wider, at its widest point, than 350mm. We paid them to work that out. They sat there, in meetings, with biscuits you and I bought, working out, to the millimetre, how big a crayfish trap should be.

They then want to know what sort of crayfish I'll be catching and, in some areas, I'll need written permission to keep them alive after they've been caught. And then after I'd waded through all the bureaucracy and the rest of the farm was wilting from my absence, I got a message saying, 'The Environment Agency is currently unable to process applications to trap crayfish.' Presumably because they're all working from home.

This means I can't even begin to unravel the rules on where the crayfish can be eaten. There's something about this only being possible on the site where they were caught but what does that mean? On the exact site? Or on the farm? I daren't ask because they'd have to have another meeting – on Zoom I imagine – and that'll cost the taxpayer another eleventy million pounds.

The upshot is that I won't repair the dam, the lake will disappear, the village will be deluged, the American crayfish will continue to wreak havoc, the voles will have nowhere to live, the ducks will bugger off and the countryside will be a little bit worse as a result.

Whereas if the government employed fewer spies and fewer bureaucrats and wrote fewer rules, it'd be a little bit better. And we'd have lower taxes.

OWL'S HOUSE

CARAVAN PARK

DRY STONE WALL

THE PIGS

ME

WASABI

MY DAM

DIDDLY

THE FARM SHOP

THE WOOD

THE SPRING

SHEEPS

JAMES MAY is A DILDO

THE CHICKENS

THE BEES

THE TROUT POND

UAT

World According To
CLARKSON

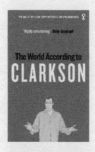

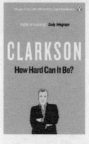

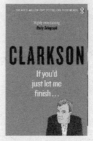

WHICH BOOK WILL YOU READ NEXT?

CLARKSON
ON CARS

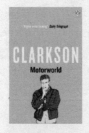
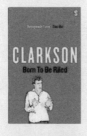
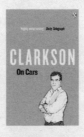
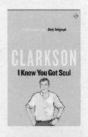

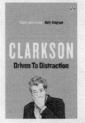
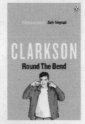
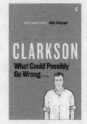

WHICH BOOK WILL YOU READ NEXT?

'Brilliant . . .
laugh out loud'

Daily Telegraph

The No. 1 *Sunday Times*
Bestseller

NURTURING WRITERS SINCE 1935

He just wanted a decent book to read ...

Not too much to ask, is it? It was in 1935 when Allen Lane, Managing Director of Bodley Head Publishers, stood on a platform at Exeter railway station looking for something good to read on his journey back to London. His choice was limited to popular magazines and poor-quality paperbacks – the same choice faced every day by the vast majority of readers, few of whom could afford hardbacks. Lane's disappointment and subsequent anger at the range of books generally available led him to found a company – and change the world.

'We believed in the existence in this country of a vast reading public for intelligent books at a low price, and staked everything on it'
Sir Allen Lane, 1902–1970, founder of Penguin Books

The quality paperback had arrived – and not just in bookshops. Lane was adamant that his Penguins should appear in chain stores and tobacconists, and should cost no more than a packet of cigarettes.

Reading habits (and cigarette prices) have changed since 1935, but Penguin still believes in publishing the best books for everybody to enjoy. We still believe that good design costs no more than bad design, and we still believe that quality books published passionately and responsibly make the world a better place.

So wherever you see the little bird – whether it's on a piece of prize-winning literary fiction or a celebrity autobiography, political tour de force or historical masterpiece, a serial-killer thriller, reference book, world classic or a piece of pure escapism – you can bet that it represents the very best that the genre has to offer.

Whatever you like to read – trust Penguin.